CSI: Momer

Forensic Science

James O. Pex

ISBN: 1983570664

ISBN 13: 9781983570667

Prologue

As a forensic scientist and as a responder to crime scenes, I witnessed some of the worst humanity could offer. But on occasion, there was a fun side to forensic science when applied to solving crimes.

A detective came to me one day and asked for some fake sodium cyanide. He was working undercover with a suspect who wanted to poison his wife but did not have any cyanide. I was aware of the appearance of sodium cyanide and needed to make up something that would look the part but would be safe to consume, should that happen.

I found a small bottle that had contained some testing sticks. It stated on the label, "not for ingestion," and had a poison icon on the label, too. I scraped off most of the label except the important words and poison icon, and then I searched the lab for the right chemical. Sodium bicarbonate in a saturated solution with water really looked the part. To a layperson, this was baking soda. I placed the suspension in the small bottle and handed it to the detective for delivery.

A few days later, the detective told me the suspect was really excited about getting the stuff and said it was the best cyanide he had ever seen. Working with the undercover detective, the husband went on to prepare a meal for his wife and place the substance in her food.

The undercover detective was invited to stay for dinner, and as soon as dinner with the fake poison was served up, the husband was arrested for conspiracy to commit murder.

You can't make this stuff up! It is amazing what lengths people will go through to commit a crime.

Table of Contents

Introduction

When a young person leaves high school, and looks to the future, the hope is that one day he or she will have a job that they will like with enough money to support themselves and perhaps a family. Some may aspire to have the job of their dreams where going to work each day was an opportunity, not a requirement.

When I left high school, no one in my family had ever attended college. Most of my friends found work in the local saw mills, but I believed the way out of poverty was with an education. At that time, the Oregon Institute of Technology was a small college that offered a medical technology curriculum. Most classes were in biology and chemistry, but the labs were designed around working in a clinical laboratory. When I graduated, the concept of like or dislike of this new career was not an option. I needed a job.

The college prepared me well for a career in a clinical laboratory and I spent ten years working in hospital labs. After the first few years I was economically stable and had the chance to look around at what else the world might offer. Working in a caste system in a hospital required a certain personality that I did not have. I was struggling and doing your best was difficult when you did not really like what you were doing.

I enjoyed the patient contact, but the lab work was monotonous. I needed to move on, this was a good job, but not the job of my dreams.

Undergraduate work got me out of poverty, college in the form of graduate school could get me out of a job I did not like and perhaps into something more interesting. With a wife, two children, and a mortgage, I went back to school full time. I continued to work 40 hours a week and finished up the degree in two years taking 12 credit hours or more each quarter. That was difficult, and I don't recommend it, but I was motivated. If you personally did not have this experience, you don't understand.

After receiving my Master of Science degree, luck would have it that I had an opportunity to go to work in the Oregon State Police crime lab.

The chapters in this book cover my personal experiences both within the crime lab and some actual casework over nearly a forty-year career. In writing this book, I have avoided giving lengthy dissertations on what happened in the commission of a crime but rather focusing on my involvement in the criminal investigation. It was my hope to explain the science of evidence examination and how it applied to the case, what I did and how I did it. Other chapters dealt with my personal life within a law enforcement agency, some personal success, some hardships.

In some chapters it may appear that I solved cases all by myself. That could not be further from the truth.

Any criminal investigation, whether for the prosecution or the defense requires a team of professionals that have an impact on the final outcome. I had the opportunity to work with some of the very best people both in the laboratory and in the community. Some of the cases in this book took years to compile all the small pieces that completed the final picture and often many people had a part in assembling those pieces. Occasionally, I discovered or developed the pieces that brought the final picture into focus. None of the cases mentioned include all of the evidence received in each case.

Criminal investigation for the prosecution or the defense is advocacy. It is wiser to admit it than wasting time believing it does not exist in you or others. As a scientist, keeping an open mind in the midst of serious advocacy can be difficult but there are ways to lessen the burden. The constant theme I used in every case when reaching a conclusion that appeared the most probable, was the concept of what else was possible. Not just considering them but listing them. This brings into play elimination theory as a working model which I cover in Chapter 5. Use of elimination theory in combination with peer review are the two processes that reduce a scientist's bias in any opinion offered. These processes were the hallmark of my career. It was not uncommon for me to gather a few people from my firm together and go over what I had found. The first question being, have I missed anything? The second question, are there other choices?

Finally, when on the stand and responding to cross examination, I may offer up the names of the other scientists who reviewed the evidence and the attorney could call them also if he so desired.

In writing this book, I worried about getting the facts exactly right as I would in writing a case report. Others may remember some facts differently or have a slightly different twist to the circumstances. That is ok. It is also possible that there are facts in a mentioned case where the information in my discovery documents was incomplete. I did the best I could with the information that I had available.

I was lucky, I found the career that I was hoping to find. My personality and the work fit together and now looking back from my retirement years, it was a good run. Most of the time it was the job of my childhood dreams. I hope you take away the following statement as my mantra since my youth. That is what this book is about.

I have always believed when faced with a problem and no immediate solution, I have opportunity.

Chapter 1 The Life Experience of a Budding Forensic Scientist

I was one of those kids who had microscopes, chemistry sets, and BB guns for my recreation. I was not an in-house kind of child. My best friend, Steve, and I were out and about every day that we could and usually unsupervised. However, being left in the house on a rainy day usually amounted to mischief. My parents were divorced; my mother worked nights and slept during the day. That left us to our own devices. If we broke something, we repaired it or hid the broken parts. Sometimes it was years before my mother noticed the difference.

Then there was the world of chemistry. I would occasionally receive a chemistry set for a special event, such as my birthday, which opened up a whole new world of mischief. The chemistry sets were designed to be inert and had instructions for simple experiments. But what was the fun in that? Everything was better with heat. Mix chemicals together in water in a test tube and place the tube over a wooden match, and it was not long before the concoction was airborne. It was a good thing my mother seldom looked up, because the ceiling in my bedroom was totally redecorated with chemical spatter.

With the Klamath Lake just a half-block away from my home, my friends and I would play all day on the banks of the lake and swim whenever we could get an adult to go with us. Some days we would collect lake water for examination under my primitive microscope. There were all kinds of wonderful creatures of unknown identification to be seen in just a few drops of water. As a child, microscopes in my hands were definitely safer than chemistry sets. We had no idea what we were seeing, so we made drawings of the organisms in the hope that someday we could identify them. This was my first look at the secret world of microorganisms.

One hot summer day, I made a new discovery. I wondered if other children realized that wooden matches fit perfectly down the barrel of a BB gun? It made a great projectile capable of starting a fire when it struck something. How cool was this? How far can you shoot a match and still have it light? I found out on the day I set the backyard on fire, much to the dismay of the local fire department. A memorable foray into firearms and ballistics.

As I grew up in a small town and came from a financially strapped family, the only science that was within my reach upon graduating from high school was medical technology at Oregon Institute of Technology (OIT). I don't think I was emotionally prepared for college, but I made the jump anyway.

It was the only college in my hometown, and the looming military draft created incentive for me to go.

The undergraduate chemistry, physics, and math classes were probably the same as anywhere else—and just as boring. What was different at OIT was the applied science associated with the clinical laboratory sciences. Parasitology, immunology, anatomy, and physiology were the heart of the program and the places where I shined. On weekends, like a lot of teenagers, my minor was in beer, girls, and good times. But school was not without a few cherished academic moments as well.

Where I first flourished was in parasitology. This was the study of microorganisms found in the gut of humans and other animals. We would study slides of trematodes and nematodes, tapeworms, and other related vermin under the microscope, which was just as fascinating as and not far from my childhood pursuits. In real life, these critters resided in the intestines of their host, so poop was the source of most of these little vixens in a clinical laboratory. What was really fun was getting our own animal specimens and bringing them into the lab for dissection. This was the exciting world of the unknown. I had lots of ideas for specimens and only needed an associate to share in the daring moments of the final collection. My friend and classmate Dale would soon to have some new life experiences hanging out with me. After all, isn't that part of why we go to college? The experience?

One day in class, we had studied how cockroaches carry a certain amoeba that can infect humans. This was really interesting, but what about getting our own samples? Did the organism exist locally? Should I add that cockroaches need to be captured alive, because if they die, the gut microorganisms deteriorate and cannot be detected? Dale and I had been duck hunting the previous weekend near a small town and stopped in at an old motel-restaurant complex for a hamburger. This place was a real dive. We looked about and quickly left to choose another location for a meal.

Now if ever there were roaches to be had, this had to be the place. I told Dale that we needed to go back, but he was apprehensive. Unlike me, Dale had a proper upbringing, and my approach to situations was often not in his comfort zone. He was a geeky kind of guy and a little introverted, especially when there was an inadequate amount of beer in his system for the fortitude required for my kind of adventure. Going for food was one thing; cockroaches at a public restaurant was another. I told him I would handle it all in the name of science. After all, if they threw us out, who would know?

We drove back to the small town, drove past the restaurant, drove by it again, and then drove by it one more time before I told Dale it was now or never; he had to find some courage, or we were going to run out of gas. He parked at a location that offered a quick getaway, and we entered the front door of the greasy spoon. We walked to the counter with our plastic jars and were asked if we needed a menu. After a short pause and a glance at Dale, who was perspiring profusely, I said no and that we were there for something else. I managed to blurt out to the owner that we were looking for cockroaches for a project in school. Dale almost fainted and was glancing at the door when I got the words out.

To our surprise, the proprietor was more than happy to help us, all in the name of science. He took us back to the kitchen, a den of smoke, grease, and sizzling French fries. The smell of grease was overwhelming; smoke was rising from the stove, and the cooks with soiled aprons were bustling about. The owner quickly pulled out a small stove, and there they were, looking back at me with beady eyes and wiggling antennae. The little brown buggers seemed docile except when I tried to catch one. Then they all took flight and were amazingly fleet of foot. Did I add that there were actually customers sitting at tables having a meal while we were in the kitchen on the hunt? I was on the floor while Dale was working a wall trying to get our hands on these speeding little bugs. We were working around the kitchen staff at the same time.

We were not having much success, having acquired only a few specimens, when the proprietor's ten-year-old son came in and asked what we were doing. His hair was unkempt, and his clothes were disheveled like he had been playing football all day. We told him what we needed, and he was excited to join in the capture. The hunt was on again. There was some daylight showing in his old and worn tennis shoes, but he was fleet of foot anyway. He said to follow him, and we went out in the main dining area where a few people were eating. We pulled out an empty booth, and there were several of the little rascals. The kid was a master at capturing them and provided us with several. By now Dale was having heart palpitations, and I could see a slight quiver in his lips as he tried to avoid the eyes of the few patrons present. Just then a roach ran under a booth where two people were eating, and the kid was in hot pursuit. The people just moved aside, and the ten year old, reaching between their legs was triumphant.

During the thrill of the hunt, the proprietor did not come out of the kitchen, and the customers did not seem alarmed. I kept telling myself, "Keep up the courage; it will work out." We had about all we needed, and Dale still had that deer-in-the-headlights look associated with that constant mental picture of where the door was located just in case this turned dark on us. As we left the building, or should I say darted from the building, Dale turned to me and stated that he was never going to follow me on another crazy adventure. These were words he would eat more times than he would want to admit in the following years.

Back at the college, we were elated to present the other students with our catch, and the lab session proceeded with the dissection of the roaches. Dale acted like the trip was really nothing and partly his idea. The professor asked where we had got them, and I said that if you ever go duck hunting near a certain small town, don't stop to eat at a certain restaurant. I'm sorry to say, we did not find the pathogenic organisms we were after, but perhaps that was a good thing for the patrons of the restaurant. Every time I would lead Dale on one of these escapades, he would tell me he was not going to follow me on another. Not until next time.

After that successful adventure, we moved up to birds as a source of specimens. The limitation was that you had to get the dead bird back to the lab in short order to get the guts out before the organisms decomposed. Dale and I wore slacks, dress shirts, and ties to school every day, and this was also our attire when hunting birds. We obtained several specimens under the college collection permits and found parasites in all of them. This was the new hunting grounds for weird species. The parasites of interest were trematodes, sometimes called *flatworms*. They inhabit the intestines of various animals, including birds.

We would collect the trematodes from the gut, mount them on slides, and then determine their identity. For two teenagers, what could be more fun? My professor was happy, and our grades reflected our efforts. We were the only students in the class interested in going out and getting our own specimens. I wanted to be a parasitologist; if I could get a career doing this, what a job. Dale felt the same way. For a couple of small-town guys, there was a whole new world out there to be explored.

One day we were out along the lake near a sawmill looking for specimens; it was spring, and the air was warm. There were log rafts in the water associated with the local sawmill, and a few ducks were swimming around nearby. I saw a Western Grebe, a diving duck common to the local area, dive and come up again. Each time the duck got closer to the log raft. I continued to watch him swim near the log raft and decided that if he got close enough, I would take a shot. I had a .22 rifle, and at about a distance of seventy-five feet, I hit it in the head on my first shot.

I handed the rifle to Dale and ran for the grebe. I jumped onto the raft of logs and ran from one log to another, getting to the outside edge where I swooped up the grebe and headed back toward shore. I was running from log to log as they bobbed on the surface. To my surprise, the logs had drifted out, and it was apparent that I was not going to make it to shore. But time was of the essence; I threw the grebe to Dale and jumped as far as I could. It was not far enough, and I fell in the lake, in my good clothes.

Covered with mud and algae, I dragged myself up on the bank, and we were headed for the college. Dale balked at me getting into his car while I was soaking wet, but what was there to do? I got in anyway, and he made offhand threats all the way back to the lab—something about all the beer that I was going to owe him and the amount of work it was going to be to clean up his car.

As we entered the lab, my shoes squeaked, and I was still soaked from my collection effort. We sat down and cleared the grebe's gut of all the little life-forms, mounted what trematodes we could find on slides with the appropriate staining technique, and finished the class. The next day we sat down at the microscopes to see what we had and set up to identify the critters with what literature we had available. To my surprise, I found a trematode that I could not ID.

I spent the next several days searching the literature, and the best I could do was get it to the genus category. I contacted the professor and told him I had one that I could not ID. He did some research as well and concluded that I may have found a new species.

I was speechless, which does not happen often. People go a lifetime in the study of small organisms and never find a new species. Here I was, nineteen years old, and I had made a scientific discovery that would persist beyond my lifetime. As is common in the college environment, my prof published a paper on the findings and signed his name to the paper.

I did not understand why I was left out until he explained to me that it was common practice. Besides who was going to believe that a nineteen-year-old would make a discovery like this? His name gave the discovery validity, not mine. What he gave me was the opportunity to explore and to research that few students have experienced at such a young age. That was my reward. It was a pivotal moment in my life; I realized that science would be my future, and it seemed all things were possible if you worked hard enough. Success does not come easily, and it was the extra effort that was the difference between success and complacency.

I have often thought that scientists could take a lesson from an auto mechanic in a small mechanic's shop. These people face problems in every vehicle that comes through the door. They need to do what it takes to get them running again with the tools they have on hand. They cannot wait for someone else to come up with a solution if they want to get paid. If you don't have the right parts, you improvise and take responsibility for solving the problem, or not.

Anyone can memorize what is in a textbook, but application requires natural curiosity combined with self-determination, the fundamental requirements of any successful scientist.

Chapter 2 A Change in Careers

After graduating from OIT in medical technology, I was living in California and taking graduate classes at San Francisco State University. I was interested in everything science and took additional classes in nuclear physics, anatomy, and physiology. New medical technologies involving radioisotopes that were impacting the medical field were especially interesting. While we were in California, my wife and I had our first child. At this point we decided to leave and return to Oregon. I enrolled in graduate classes at Southern Oregon University to finish up my master's degree, which included more biochemistry, graduate-level physiology, and other courses leading to a minor in education. By this time, we had two children, a mortgage, car payments, and the usual responsibilities of a young family. In order to get through school, I worked nights at the hospital and went to school days, carrying at least twelve credits per semester.

Late one afternoon while I was working in the blood bank section of the lab, two guys from the local crime lab came in to get some blood from each ABO blood group. The ABO blood groups are the most common blood types used for transfusion. Everyone is either A, B, AB or type O. Blood units have small segments attached to them that contain the donated blood.

These segments can be used for testing to ensure that the units are compatible with the recipient without getting into the unit itself. My undergraduate studies included immunology, the study of antibodies and antigens in the body as well as methods for their detection. This included ferreting out antibodies in patients that could cause problems when the units were used the next day for surgery.

I was curious why these guys needed the blood, and we talked. They had some new procedures from the FBI for determining the ABO blood type in dried blood. All the work I had done was with liquid blood; there was a difference. I had no idea what people in crime labs did, but it was apparent that these two were chemists with no background in blood typing. In the following weeks, they came back several times to get more blood. The procedure they were provided was not working, and they had no idea why. So I told them to bring me the procedure, and I would take a look at it when I had some time later in the evening. It did not take but a moment to realize why the tests did not work. The crux of the problem was the temperature the tests were run at and the way that the bloodstains were fixed to the cloth. One step in the procedure called for drying the reconstituted bloodstain on a glass plate. Then the plate was boiled in water! I had never heard of such a process and had no idea why someone would do that.

From an immunology perspective, this made no sense. After making modifications, I was able to make the tests work and called them back to the lab to see what I had done. These were nice people, but it was apparent that they had no idea what I was talking about. Terms such as *macroglobulins*, *antigens*, and *antibodies* were not part of a chemist's vocabulary. They were polite and left with the new information. Despite the clarification, I knew when they walked out the door they were not going to be successful with the procedure

The next day I got a call from the director of the crime lab system under the Oregon State Police seeking information about me and eventually getting around to offering me a job. Coincidentally I was looking for a change. Some jobs require a certain personality and working as a medical technologist was a misfit for me. I just could not get used to the caste system common within the hospital environment. I needed out. I had recently taken the national exam in advanced clinical chemistry. I passed in the top ten percent in the country and thought this was something special. My boss purchased some doughnuts and wished me well and then advised I get back to work. Instead of being upset for the lack of acknowledgment for what I had accomplished, I thanked him and went back to work. It was the message that this profession was not for me. He helped me to see it. The next day I called the crime-lab director back and we talked.

I traveled to Eugene for an interview and was advised there was only one position available and there were other candidates also being considered. While I was there, an analyst had a woman's thigh on a counter top. Her thigh had been found in a dumpster. No one had any idea how to get a blood type or where to collect a sample for that testing.

Dead folks did not bother me. I used to attend autopsies and do cleanup for extra money when I was an undergrad. With their permission, I picked up some swabs and collected blood from the bone marrow. I asked them if they wanted me to type it for them while I was there. Needless to say, I made my point, and I got the job. In the late 1970s, I could not believe they could not do this basic blood grouping in a criminal case.

During the interview, they wanted to know if I was familiar with firearms and cameras. I had owned several firearms most of my adult life and was an active deer and elk hunter, so no problem there. I used to do wedding photography as a hobby with my own SLR camera for extra money—again not a problem. So I asked them if I could see

their cameras. There were two types of cameras available. The "modern" cameras were Canon F-1's. They were among the first to have internal light meters, but you still had to adjust the shutter speed and the lens opening (f-stop) to get the right exposure.
This was several generations behind my personal camera. At least they had recently moved beyond flashbulbs and had a decent flash setup.

The other cameras available were Graflex 4x5s, kind of like what you see in old movies. With the old Graflex, one had to use a light meter to determine the right combination of shutter speed and f-stop and then make the adjustments to the camera. Then a film back containing two negatives was placed in the back of the camera. A blank slide was moved from in front of the film to keep out light. The shutter was cocked, the slide was opened, and the film was exposed by releasing the shutter. The slide was replaced over the exposed film; the back was removed and turned over, and then the second film could be exposed in the same manner described. The exposed film negatives protected in the back were now ready for developing in a dark room where the film was manually processed into an image. The crime lab had a dark room for just this purpose.

We had several discussions, and I agreed to take the position in the crime lab in Eugene, Oregon. However, I was advised that the superintendent of state police had the final say on my duty station. Nothing was ever guaranteed. The pay was not very good, but I needed a change. In fact, the pay was so poor that my family qualified for low-income housing. After eight weeks of recruit school, I did get the transfer to Eugene, and my life began anew as a criminalist and a sworn police officer. In the minds of the OSP, this was the lowest form of law enforcement, and we were called *technicians* internally.

Processing homicide scenes and court testimony as an expert witness in a death-penalty case was so far removed from upper management that they had no idea of our value to the criminal-justice system, their focus was on traffic. When I was hired, there were only five of us in the lab. I was impressed by the professionalism of the folks and their dedication to their jobs. It was not easy to gain a position as an officer with the Oregon State Police and also to meet the science standards for the crime lab. All the people I worked with in the early days were special in my eyes.

My first days on the job were bewildering. All training was on the job initially, and we often encountered problems for which there was no immediate solution. What was important from the first day was how you reached decisions. You followed a fact trail and opined on a fact-based conclusion. As cases became more complex, we became experts in scientific method and the procedural steps involved. Often the final step still allowed for several choices or a *differential diagnosis*, as the term was used in medicine. What was different in our profession was that if you cannot exclude a possibility, it remained possible until you can. In doing a hair comparison, for example, there may be certainty in hair that you can exclude. There remains uncertainty if you cannot exclude.

In the late 1970s, forensic science in Oregon was just getting on its feet, and there were often unanswered questions. Reliance on our own experimentation often was the best proof. We often took the time to experiment with bloodstain patterns to see what we could learn and also to explain what we saw at a crime scene. One day my boss, Chuck, and I were out in the garage setting up a shooting experiment with a sound-trigger flash to see if we could capture the blood in flight. A bloody sponge was placed near the floor with white papers behind, and the camera was ready to go.

I stood back as Chuck touched off the round, shooting at the bloody sponge with a .22 revolver. The bullet passed through the target, struck the floor and, then the refrigerator, and came to a stop, hitting me in the chest. There was no penetration, but what an opportunity. I quickly fell to the floor and claimed that I had been shot. Chuck was beside himself and in a panic. I struggled on the floor and moaned as if it was my last. Chuck ran to me, rolled me over, and was full of sorrow—or so he stated—until he saw the grin on my face. I just could not hide it! What fun! To this day, I am probably the only forensic scientist who had been shot by his or her boss.

Most of our days early in my career were filled with doing drug cases, since this was the most common evidence to come through the door. The analysis of drugs at that time was done by color tests, which were presumptive crystal tests, and basic instrumentation, such as a gas chromatograph. A gas chromatograph, at that time, was the most advanced instrument in the laboratory. The instrument was operated by injecting a liquid extract of a powder into a heated chamber that vaporized the liquid. This was under pressure, and the drug was forced into a coiled steel tube filled with special reagent that acted like a sieve. Compounds separated as they passed through the column. At the end of the column was a detector that would quantitate the presence of any organic compound passing by. The time it took the drug to pass through the column was the measurement used for identification as compared to a standard.

There was no money to pay for repairs, so we had to repair these instruments ourselves. When the instrument started plugging up, this meant we had to drain the old solid reagent out of the column and force new reagent in by attaching a funnel containing the reagent at the top and applying a vibrator to the side of column to jiggle the reagent into the tube. I look back on this and laugh at our "advanced" methods compared to what is available in a modern crime lab today.

Crystal tests were interesting, in that you took a bit of the suspected drug, did an acid-base extraction, and dried the drug onto a microscope slide. Then a special reagent was added, and a unique crystal shape formed in the chemical reaction. Unlike a snowflake, the shape was consistent and reproducible for a specific drug. Unfortunately, it required a pure sample to get the diagnostic crystals. That was not always the case, and there was no way of documenting your observations other than just stating in your notes that the test was specific for the expected crystals.

Occasionally some drugs had to be tested by thin-layer chromatography (TLC). In this process, a possible drug was diluted in alcohol and dried as a spot on a glass plate, which was coated with a special compound that allowed liquids to advance up the plate when the end of the plate was inserted into a liquid chemical mixture. The separation of various drugs was dependent on molecular weight and solubility in water, also called *polarity*. If you take a sheet of newspaper and dip one end in rubbing alcohol, the water will wick up the paper, and the ink will begin to flow and follow the alcohol front. This was basically the same principle.

At times we did TLC with filter paper. Besides drugs, if someone forged a signature on a check or other official document, we would extract a very small sample of the ink and compare it to a seized pen or other inks on a document by TLC. It was an excellent tool and fun to run the test. A single black ink could generate ten sub-colors, which would separate out along the path to the top of the plate. Then we could change the solvents and get an entirely different pattern if we needed better proof of an altered document.

A bad check might be sent to the lab to determine if the ink was from a certain pen or if the number of zeros after the first numbers of an amount on the check all came from the same pen.

As generalists, we learned all sorts of test procedures. Another test was the identification of a typewriter to printed letters on a document. With a typewriter, the keys wear over time, and some of the loops begin to fill with debris. These characteristics were available for comparison, but if this was not useful enough, one could also unwind the ribbon and look for the sequence of letters that corresponded with the evidence. This was considered a tool-marks exam and comparing a typewriter to letters on a document was not that unusual for us to work. It may seem a stretch for a chemist today to do such an exam, but in those days, there was no one else. If we did not do them, the test was not done.

Many years later in the age of computers, a reporter in Israel contacted me about examining President Obama's birth certificate to determine if it was fake. He sent me a digitized file of the document. Since I had done some documents work, the first thing I noticed was that it had been completed with a typewriter and not a printer. I could see defects in the stamped letters of the document that might be useful if a typewriter was located for comparison.

I passed that information along, but no typewriter was ever located to my knowledge. I presumed that after the statement was made, the issue was dead. Then a few years later, I heard that a right-wing talk show host in the Midwest stated that I had proved the birth certificate was a phony. You wonder about the relationship between the Israeli reporter and the Midwest newscaster, but apparently this was how fake news gets its start.

Running tests in the lab was part of the job, but meeting with detectives and going to scenes was the other part. We normally did not wear our uniforms in the lab, but we often carried our badges and firearms when out of the lab. I think people working in crime labs today would be surprised knowing that we carried .357 Magnum revolvers with six-inch barrels under our sport coats. Sometime later we were issued

.38 Special revolvers with two-inch barrels, which made concealment a little easier.

There was a problem in that we had to qualify at the range with both weapons every ninety days. This was not a problem with the .357 with the longer barrel, but the snub nose was much less accurate. In fact, I could not hit the broad side of a barn with it. In qualifying on the twenty-five-yard shooting course, it started with shooting a selected number of rounds with each hand at twenty-five yards. As you got closer to the target, a selected number of rounds were placed in each firearm and discharged on command. Based on the initial appearance of my targets, missing the correct number of bullet holes, I needed to do something.

Usually there were several officers qualifying at the same time, and the number of shots individually fired at each station was not noticed by the range commander. So I adjusted the number of rounds. Fewer rounds at twenty-five yards and more rounds at seven yards. When the final tally was done on my target, I qualified. Did I mention that innovation was required at times in this new job? Failure was not an option. I did well with the .357 anyway. But with the .38 snub nose, there could be an army of terrorists out there at twenty-five yards, and they would be safe with that firearm in my hand. They were in more danger from me throwing rocks than shooting at them with that snub nose.

Chapter 3 The State of Forensic Science in 1978 in Oregon

The 1970s marked an era in which forensic science was in its infancy in Oregon and labs were initially staffed by a few doctors or other medical people at the University of Oregon Health Science University. Testing was limited to toxicology, blood alcohols, and some evidence examination.

The Oregon State Police was divided into several divisions: game enforcement, traffic, criminal, and the labs. A real crime lab was established for the first time in Portland and staffed by police officers with science backgrounds. In the mid-1970s, the lab system expanded to include a laboratory in Eugene and Medford as well. Just prior to my hiring in 1978, the OSP received a federal grant that allowed the expansion of labs to Coos Bay, Pendleton, and Ontario. These small satellite facilities were two-scientist labs and designed to cover the more rural regions of the state.

The purpose of each crime lab was primarily drug testing, physical-evidence exams, and call-outs, especially homicides. OSP detectives were required to carry a print kit that contained ink, brushes, and fingerprint powder. Their skill at getting a latent print varied considerably.

OSP developed a lab in Salem for documents exams, fingerprinting, and record keeping to improve these services.

The official concept at that time was that you were a police officer with special duties that could be taught to anyone. Therefore, you were not performing anything extraordinary, and your job was considered the lowest position in the hierarchy of importance to the OSP. The upper management in headquarters was mostly traffic officers with no college education, and to make matters worse, they considered those who had degrees as overqualified and to have wasted their time in school.

Hiring within the labs required meeting the requirements for an OSP officer with a college degree in one of the sciences. Despite our low status with the department, our hiring requirements were more stringent than any other department.

If you passed the initial entrance requirement for the Oregon State Police, you went to Salem for interviews, and if you passed, you were assigned to the next recruit school. Without a doubt, this was a paramilitary organization. Recruit school was eight weeks at an air base in eastern Oregon and reminiscent of my basic training at Fort Ord. Physical fitness, shooting, and classroom time related to the state and federal statutes filled the day. We were trained in interview techniques, use of the radio, driving patrol cars, DUII stops, and making felony stops. Training in deadly use of force was especially important.

We were armed with heavy .357 Magnum revolvers, and in shooting exercises, we would fire up to twelve hundred rounds a day. I was thirty years old and one of the oldest in the academy, but as I could still shoot and was in good physical condition, I did OK. Despite my academic credentials, I cannot spell worth beans. The camp staff were more than happy to remind me of my spelling errors, suggesting that my education was probably a waste of time. My father used to tell me to picture the word in my mind and just read off the letters. The theory was good, but I lost something in the application. Any time you did not meet the standards set forth in the school, you were discharged and sent home that day. In my class, some did not make the grade. In addition, upon completion of the school, the new recruits were assigned to offices at the direction of the superintendent, and nothing was guaranteed. I was often told that I could be assigned to traffic if there was a need. Fortunately, I was assigned to the crime lab in Eugene.

In addition to our assignment as OSP troopers, we were also called *criminalists*. We were trained in many of the specialty fields, and as a result, we were also called *generalists*.

Forensic science is the application of science to matters of law. We ran tests in the lab, attended crime scenes, attended autopsies, prepared media for trials, educated police and prosecutors on all aspects of forensic science, and kept up with the sworn officer responsibilities with special training. Our most important responsibility was testifying as an expert witness in a court of law. We applied our trade to matters of law.

A true definition of our role: We were police investigators with a science background. We worked with detectives, attended team meetings with agencies, and applied science when applicable. The chapters to follow in this book will demonstrate the application of science to matters of law.

Most of the training was in-house, and the *Journal of Forensic Sciences* was the only outside resource available. In the beginning, I took home a journal every night. My goal was to read every article published in the journal back five years.

It took several months to accomplish, but at least I understood the state of the forensic sciences when I finished the oldest journal.

In spite of the broad spectrum of training, people tended to migrate to specific areas of interest and could be relied upon to provide advanced skills if you had a problem. We were each assigned a county within our lab region and were primary in handling all forensic issues associated with that county. We could wear our uniforms any time, but most days we were in lab coats. When we went out on call-outs, we had jump suits complete with our badge and weapon identifying us as police officers.

At times when I encountered something new while working a case, I would research it and experiment until I understood the science. If I did not have enough time at work to expand my knowledge of a specific science, I took the papers home and spent additional hours learning on my own time.

Forensic science in the 1980s was growing fast with new tests, new instrumentation, and new procedures. With forensic science highlighted in some high-profile cases, the community was beginning to recognize our contributions to criminal investigation. The FBI recognized the lack of training across the country and hosted a number of specialty classes to bring the skill levels up in the state and local labs. All of us attended a school in Quantico, Virginia, at least once a year. After one of us attended a school, we held an in-service, and the trainee was expected to pass their new knowledge.

The department would pay for a criminalist to belong to one forensic organization, and you may get to attend a forensic meeting once every few years at department expense. To get to the meetings with any frequency, it was not uncommon for us to group together and share rides and motel rooms so that we could get to more meetings on our own funds.

Keep in mind that we had young families, and my wages were low enough that I qualified for low-income housing. Getting to a meeting on your own hook was a big deal.

It did help getting the time to go to a meeting if you could present a research paper. So I would develop some new techniques in blood typing or present a unique case in order to get to additional meetings. In 1984, I presented the Diane Downs case at a meeting of the Northwest Association of Forensic Scientists. To my surprise, I was selected to receive a special award from the American Academy of Forensic Sciences for outstanding contributions to forensic science the following spring. Being honored in front of fifteen-hundred of my peers was one of the highlights of my career.

The professional nature of all the scientists that I encountered within the Oregon crime lab system at that time was above reproach. I think it was the need to pass the combined requirements for an OSP officer and the requirements for the crime lab that brought out the best people. The science background of the personnel varied from person to person, as that was not the primary consideration for employment. An advanced degree was nice, but the ability to communicate was usually the deciding factor in hiring. No matter what information you had for a jury, if you could not get it across to them, your presence was a waste of time. A criminalist also had to have a personality that would withstand the rigors of the job as well as strength on the witness stand. Almost everyone was in it for the career; thus, turnover was negligible, and new hires were usually based on retirements or expansion.

Today the crime labs are highly specialized, in that scientists seldom have more than one specialty. The concept is that they will become more proficient at doing one thing every day. They are no longer sworn officers; they do not carry weapons; and they do not attend investigator's meetings but remain in the lab, except for those trained to do call-outs.

Today crime laboratory scientists are in the business of running tests. The positions can be tedious, and personnel turnover is much higher than it used to be. The last new hire I mentored wanted overtime if she took material home to review. What was this? Where was the instinct for knowledge and the drive to be the best? If she only knew the esprit de corps that existed in times past, perhaps she would want to meet a higher standard.

In those early days, we were the bridge in all matters pertaining to science in its relationship with the law.

Chapter 4 The Beginning of a New Adventure

The first few years in the lab in the late 1970s and early 1980s were hectic with a heavy workload and frequent training. Chuck, the lab director, was my coach and a very good scientist. I owe much to him for what he taught me in those early years. There were five of us in the lab to cover a large segment of the Willamette Valley and south as far as Roseburg. Terry Bekkedahl, Linton VonBeroldingen, and Kenn Meneely were the other three scientists. All five of us in the lab were Oregon State Police officers and forensic scientists and were required to meet the continuing training for both professions. For the first time, I felt like I was in a professional work environment and not just a job.

The first crime scene I attended involved a woman who crawled out of a ditch alongside a busy highway and flagged down a passing vehicle. She claimed she had been assaulted and raped by three Native Americans in a black Trans Am. They reportedly had struck her on the head with a rock, ripped her clothing, assaulted her, and left her at the roadside. She reported that her mother-in-law was missing and also reported the location of her truck farther down the road. A search for the mother-in-law was initiated, and she was found dead at a wayside a few miles away from where the reporting person had been abandoned. The mother-in-law had been shot five times.

Chuck and I traveled to the scene, met with the detectives, and began processing the area. We did not find much, but there were some tire tracks, footprints, and miscellaneous papers that may have come from the get-away vehicle. There was also a short length of steel wire called bailing wire. The wire was on top of the leaves and pine needles, which meant that its deposit was recent. We placed placards at the location of each item, photographed them, and were in the process of placing the items in bags for return to the lab.

While we did our processing, a news helicopter flew over the scene at close range. Then everything went nuts; the wind from the rotor quickly picked up all the evidence and redistributed it about the area. Everyone was yelling at the helicopter crew, and as I chased my evidence around the scene, I had the overwhelming desire to take a few shots at the helicopter. They quickly left, and we resumed our collection of what evidence we could find. After the body was removed for autopsy, we dug up the underlying dirt and found four bullets that had passed through her body and into the ground. She had been executed at that location while on the ground.

We took the evidence back to the lab, logged it in, and then processed all of it ourselves. Reports were generated later, sans the helicopter experience. On close examination of the daughter's clothing, we found that there were several rips in the blouse, all vertical and all parallel. This supported her story that her clothing was ripped when she was raped. On microscopic exam, it was clear that something was amiss. Each rip began with a sharp cut to expedite the parting of the cloth. Something was wrong here; the rips now looked staged.

Further searching of her clothing revealed a note in her coat pocket with some numbers that we suspected were milepost numbers on the highway where she was found. Chuck and I went out to that location, started sifting the soil with a rake, and found a partially buried plastic bag. Inside was a .357 Magnum revolver with a cylinder capacity of six rounds. One live round was still in the cylinder. Using a comparison microscope, Chuck compared the bullets from the ground that was under the body of the mother-in-law to test-fired bullets from the handgun and determined that they matched. The bullets from the scene had been fired through the barrel of the revolver found along the highway.

The daughter-in-law did what some people sometimes wished they could do; she offed her mother-in-law. She just did not count on us being as thorough as we were in our investigation. We also discovered that daughter often carried hay for her horses in the pickup, and the wire found at the scene had the same class characteristics as the wire found loose in the pickup, hay-bailing wire. The wire at the scene just did not belong there; we knew that it might be important but did not know why until later in the case. I would learn that this was common to crime scenes; we knew what was evidence, in that it was out of place, but we would not know why until later in the case. Proof that good science might begin with a hunch.

Speculation based on a six-round capacity cylinder suggested that she shot mother-in-law once while she was standing and then shot her four more times on the ground. She apparently was angry. You think? Afterward she faked the assault on herself to cover the commission of the crime. Unfortunately for her, good forensics was her undoing.

Working in the lab was not without a little humor. Our secretary, Martha, had an interesting quirk in that when she needed to talk to you, she had to leave her desk and make visual contact to give you the message. It was not long before I started exploiting this behavior. I would hear the phone ring. Martha would answer, and I would hear her say, "Jim is here." Her chair would slide. She would call my name and would be on her way to seek my location. I would quickly move to another spot and wait for her to call, "Jim?" "Over here!" I would shout, and then I would move to a new location. We would keep this up until she finally caught me and I got the message. A guy has got to do what he has got to do when tied down in monotonous drug testing.

Then there was the day Kenn asked Mike to look at some evidence under the binocular microscope. Mike was unaware that Kenn had placed black fingerprint ink on the ocular cups. Mike looked into the scope, was satisfied with his findings, got up, and walked around the lab unaware of two black rings around his eyes. Every time we looked at him, we laughed ourselves silly, but with no mirror in the lab, Mike could not figure out what had tickled us so much. He finally went to the restroom and discovered the problem. It was a good day in the life of a criminalist.

The photo below, Figure 1, was of the original crew on a dress up day. I am sure just seeing this photo would instill fear in the hearts of any criminal. I think we used it on our Christmas card that year.

Top L to R:

Terry Bekkedahl
Kenn Meneely
Chuck Vaghan

Lower row:

Jim Pex
Martha West
Linton Von Beroldingen

Figure 1

Our laboratory was in an office building, and we always incurred the interest of the other building patrons. One day we decided to have an open lab day and invited some of the other folks into the lab for a tour. Before the tour began, we had a Eugene Police Department detective come over; remove his shirt, shoes, and socks; and lay out on the table. We covered him with a sheet and placed a wire tag on his toe, which was exposed. We dumped some blood in a pan, threw some on the sheet, and placed a few bloody knives on the table next to him. When the tour started, the visitors could see the toe and were quite excited.

We led them around the lab and next to the table. Most were nervous and asked many questions about how we got the body into the building without being seen. I told them it was easy; we dismembered them out in the van and just brought in the parts we needed. "Oh my God!" was heard all around the room, but the best was saved for last. We moved the folks close for the unveiling, and then suddenly the detective sat up on the table and gave a huge sigh. At that moment, there were screams, and one gal nearly fainted dead away. That moment was one of the highlights of my early career. Who says this job can't be fun once in a while?

During those first years, we were on a monthly salary, and there was no extra compensation for call-outs. The five of us were responsible to a five-county area of the Willamette Valley with a half-million population. Two of us went out on each call, so we were on call two weeks out of the month. The call-outs were often every few days, and sometimes they held one scene and waited for us to finish up another one. We went out on burglaries, sexual assaults, suicides, and homicides at the whim of management.

Some of the cases were complex and required a scientist to examine them. Others, like a dead guy in a pickup at the beach, were a total waste of time. In that case, I drove for two hours, opened the door, removed the suicide note from the dash, and give it to the detective. The deceased still had the gun in his hand. Since we were free, why not have the lab do it? That was the central theme. Holidays were the worst; it was hard to get a Christmas at home because this was the suicide season.

In the early 1980s, Pierce Brooks, a former LA detective who was popularized in the movie *Onion Fields*, moved to Eugene and was hired as the chief of police for Eugene PD. Pierce wrote a book titled *Officer Down, Code Three*. The book was one of the first on how officers need to protect themselves, and it discussed several fatal police shootings. The book covered not only an officer's death but the circumstances that were common to the situation. He called them the ten deadly errors that often contributed to the officer's death. A copy of this book could be found in nearly every police station in the country. Pierce Brooks was well known in law enforcement communities nationwide.

A few years later, he left that position at the PD and volunteered with the DA's office. When we met, we hit it off, and he often went with me on call-outs.

I always appreciated his input from years of experience with LA PD, and when we finished a scene with very little helpful information, he offered encouragement and would often comment on some esoteric item of evidence, "I don't know what this means right now, but in time it will be important." He was right more times than not. Pierce went on to assist the FBI in setting up the Violent Criminal Apprehension Program, or VICAP, which gathers information on interstate crimes. With the assistance of behavioral scientists, VICAP looked for commonality that could assist local agencies in solving crime. I miss those times on call-outs with him; we did well together.

One downside of the position was that all of us were struggling with the low salary. A Christmas in this new job was coming up, and we were so poor that it was difficult to make ends meet, despite the number of hours I was working. My mother, who lived in Idaho, decided she would sponsor Christmas and provide all the gifts for my family. My wife and I did not purchase any presents and focused on getting the family to church as preparation for the holidays. On December 23, we drove all day and purchased a room in a cheap hotel for the night on our way to Idaho. We would make Grandma's by late afternoon the following day.

That night my beeper went off, and I called the number listed. It was the deputy DA in a local county telling me I had to be there the day after Christmas for trial—no exceptions. This was a crushing defeat for my wife and me; we were not able to proceed on to Grandma's, and we had nothing but some holiday cookies for the children. The children were asleep.

We put our heads in our hands and mourned our predicament; at least we had the cookies. Early the next morning the kids got the bad news. We were all heart broken. We had no money for gifts, and we had to return home. I walked over to the cookies to snatch one for a breakfast treat and discovered they were covered with ants!

For my wife, this was the last straw; she broke down, and we cursed this decision to leave my previous profession for this one. As we sat there in our misery, I was reminded of my father's words for just such an occasion: *When the going gets tough, the tough get going.* It was December 24, and I told my wife we were going to go home and make a Christmas for our children, no matter what. We drove to the house. She set about cooking new Christmas treats as I went about maxing out the credit cards to put gifts under the tree. The children thought Christmas was wonderful, and that was all that mattered. After Christmas, I testified in the trial, and we moved on. My wife and I sometimes reflect on this low point in our life and the way we attacked adversity in the worst of times.

After I was a few years in the new position, the work was beginning to show a downside. The stress, the long hours, and the deplorable circumstances often associated with a serious crime was starting to get to me. In our capacity as responders to crime scenes, we witnessed some of the worst humanity could offer in the destruction of each other. We applied our science, collected the evidence, went home to our families, and tried to smile. I could tell by the look in my wife's eyes that some days it just was not working.

I could no longer remember the names of the dead, because there had been so many. The fractured skulls of the children, the dismemberment of the women, and the shotgun blasts to a person's head were beginning to add up in my brain. This was compounded by having to attend every autopsy in each case as well. The stress was getting to my associates as well, and I could see it while they were in the lab. Some days they just sat there in silence, unable to get anything done, not knowing what to do next. I felt sorry for them but wondered what they thought of me. Was it showing in my day to day actions, too? What was happening to us? We used to be so excited about this job.

In our business, you try to focus on the science at the crime scenes and not let your mind wonder about this person or persons who were now dead on the floor. That was the theory anyway. I'm not sure we ever overcame this entirely, especially when we were being sent out to pick up a newborn child abandoned at a roadside park who died of hypothermia or a young child strangled by the mother's boyfriend. In some of these cases, there was just not enough science to overcome the emotions, despite my best efforts. My friends outside law enforcement and our marriage-encounter group at church kept me in touch with the societal norms and continued to remind me that what I saw in my work was not normal. My non-law enforcement friends continued to be a blessing throughout my career.

At a division meeting of all the commanders, a few of us stood up and addressed this problem with the superiors at headquarters. We got an immediate response: "Buck up, or get a job somewhere else." Some did. As we were told in recruit school, we served at the pleasure of the superintendent. He did not need much of a reason to fire us. Failure to go when called out again was a violation of a direct order, so most of us continued on. Eventually the troopers as a group decided to form a union in the mid-1980s, which was a turning point in the ridiculous call-outs. What a surprise; when money was involved, they did not need an expert at a scene to tell them it was a suicide when the gun was still in the hand of the deceased.

Today law enforcement people often have department counselors they can rely on in tough times. Early in my career, either you powered through it or you moved on to another profession. Finally, the department hired a former Catholic priest as a state trooper. He attended recruit school and later offered aid to the troops in hard times.

We still referred to him as Father Patrick, and when he was in town, he always came by to see me. We hit it off right away and we always took time for a cup of coffee. In retrospect, I wonder if he just liked to talk, or if he saw me as a client. I guess I will never know. It probably just depended on the day.

In those early days before the police union, our first opportunity for extra pay came when we were approved for seventy-five dollars for twenty additional hours of work. With our schedules, it normally did not take long to make the big bucks. But when it was occasionally slow, I made the extra money by driving for the SWAT team of the local PD. We had an old beat-up van with no side windows that we used for callouts, and it was so ugly that no one would suspect it was a state-police vehicle. I think OSP got it through a seizure. I would put on my uniform, drive to the local PD, and load the

SWAT people in the back of the van. We would be off to do a raid or, as it is properly called, the execution of a search warrant. I would drive right up on the lawn of a residence; the doors would fly open, and everyone would move toward the front and back doors. Being well educated, I knew the best position for such a raid was the last person through the door. It was often so exciting; I had to pinch myself to believe I was actually getting paid for this.

When we were working with new deputy DAs, we liked to pull their chain a little bit. We were in a pretrial meeting talking about the execution of a search warrant, and I offered entry procedure for knock and announce. I said, "Basically you just walk up to the front door, announce yourself, and ask if we could come in." Sometimes it worked, and the suspects let you in. But as sometimes it didn't work, and the residents did not come to the door. We told him we had an alternative plan. The guy at the front door would shout, "Police," and the officer at the back door would shout, "Come in!" We never really did this, but it was a fun way to get new deputy DAs to absolutely fall out of their chair. These were the days when we were both police officers and scientists—and also just human.

As the first years went by, I continued on a steep learning curve. Usually I learned within the lab, but at times it was in special schools on firearms, crime scenes, and trace evidence. I also continued going out on crime scenes on a regular basis. One scene stuck in my mind as unusual, since this was probably the first one that I did on my own. This was a case where the husband claimed his wife attempted suicide with a revolver.

The husband reportedly walked into the bedroom and saw his wife standing in a corner of the bedroom holding a .357 Magnum revolver preparing to shoot herself.

He said he rushed at her and grabbed the weapon. According to him, they struggled with the weapon, and it went off. The bullet struck her just above the bridge of the nose and exited the top of her head. She fell to the floor unconscious, and he called 911. A deputy sheriff responded, but no evidence was collected or anything unusual noted. The deputy did not have much experience with shooting scenes; he listened to the husband and determined that his account of what happened was believable.

His wife was transported to the hospital, where she underwent immediate surgery. The surgery was able to repair some of the damage, and she hung on for a couple months. Bacterial infection finally set in, and she passed due to the infection without gaining consciousness.

After her death, a detective from the sheriff's office came to the lab and asked for me. He just did not feel right about the incident, and it was worrying him. We sat down over a cup of coffee, and he showed me photographs of the scene. The photos were limited in scope, and I initially did not see anything out of the ordinary. It had been three months since the incident; I was not sure what he expected from the scene by this time.

The detective had in his possession her wedding ring, which had been removed at the hospital. The ring finger was damaged, believed to be from the bullet.

I took the ring from him and placed in under a field scope to look at the damage. There was a dark smudge in the dent that did not belong there. I was concerned that it could be lead, since there was already a suspicion that the bullet had struck her hand. I placed the ring in a special instrument and tested the ring for its elemental composition.

This area of the ring was positive for lead; other areas were not. This was confirmation that the bullet had struck the ring. The damage was on the underside of the ring, so her finger was either near the muzzle at discharge, or if the ring was worn properly, the bullet had struck her finger while her hand was in an open position. We had no information that there were any dark-stained areas on her hand from a near-contact shot, but knowing the hospital environment, I doubt anyone was looking.

This, and other information, was enough for the detective to get a search warrant for the residence. The next morning the other guys in the lab were busy with other responsibilities, and I was on my own. I met the detectives and the deputy DA at the scene. To our surprise, it had not been cleaned up. The husband had left the house and had only returned to get his clothing. Although we did a cursory search of the entire residence, the bedroom was the location of interest. The room was full of furniture, including a king-size water bed, dressers, and nightstands. There was not a lot of exposed carpet to walk within the room.

Before I was going to walk about within the room, I decided to use our new forensic vacuum sweep on the floor for gunpowder. We had the device of only a few weeks, and I was anxious to try it out. I used string and tape to lay out the exposed floor of the room in one-foot squares and vacuumed each square carefully. Then I replaced a capture dish inside for the next square. I made a diagram of the floor and numbered each dish corresponding to the appropriate square.

In the southeast corner near the head of the bed, I could see a bloodspatter pattern on the wall. The top of the pattern was about three feet above the carpet, and the center of the pattern was yet to be determined. There was a dent in the wallboard, and the bullet was located on the floor, not far away. It looked like the dent was a bullet strike, but I was not sure yet.

After the vacuuming, I got down on the floor with a bright light and did a closer examination. There is an old saying that you never know what you might find but will recognize it when you see it. Here was the true application of that statement. The waterbed frame board extended over the wooden base, which was recessed. The overhang was about six inches, and there was bloodspatter on both the base and the upper frame. Because of the overhanging bed frame, this spatter on the base could not get there from an origin higher than the spatter pattern on the wall. It was clear that the origin of this spatter was close to the floor.

Several photographs were taken, and it was time to do some measurements. Bloodspatter droplets were selected from the wall and the bed; they were numbered and photographed. From these droplets, I measured the length and width of each one, did a mathematical calculation, and was able to determine the angle of incidence for each drop selected. An example is shown below as Figure 2.

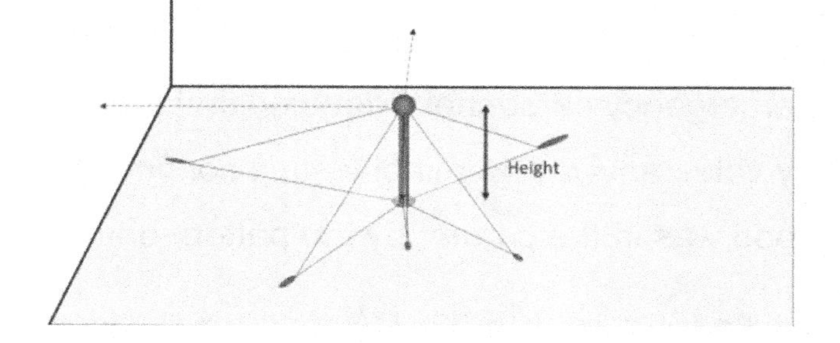

Figure 2

For the pattern on the wall, the origin was six to eight inches above the floor. This was confirmed by the angle of blood travel that would allow spatter to get under the overhanging frame and strike the base.

After I had sufficient photos, I took samples of the blood for typing purposes. The blood was the wife's blood type.

The last step was to luminol the room. In Oregon, you cannot serve a search warrant and then leave the residence and return later without having to get another warrant. It was the middle of the day, so the detective purchased several black- plastic tarps, and we hung them in the room, creating close-to-total darkness. This was required in order see the luminol reaction if it located possible blood. When I sprayed the carpet, I could see shoe impressions on the carpet but could not get a good photo of them. With the old cameras we had at the time, a single exposure took one to two minutes, lots of time that usually ended with the pattern washed out or blurry. With the number of people who had been in the house, such as emergency personnel, I doubted that the prints would be of any value anyway. Luminol testing confirmed that most of the blood was in the corner, by the pattern on the wall.

I talked it over with the DA, and we decided that the bed frame, floor, and wall were important for prosecution of this case. The DA called the county and had a carpenter respond to the scene. We drained the waterbed, dismantled it, and took the frame with us. But that was not all. He wanted the floor between the bed and the pattern on the wall also, so that is what happened. The carpenter cut out the floor and the wall with the pattern, too.

I stood by to make sure that the evidence was not damaged in the process. In the end, we left with about a quarter of the bedroom. But where to store this evidence? There was no room at the sheriff's office for this stuff. Fortunately, the detective found an empty room in the county courthouse, and the evidence was reassembled there. Oregon law required that the DA had to pay for any damage incurred, so the original bedroom was eventually repaired after we left.

Back at the lab, I took out the vacuum sweeps and examined the contents under a microscope. Fragments of gunpowder that were partially burned were found and chemically tested as well. Just going by the squares in my diagram that had gunpowder, the only carpeted areas positive for gunpowder were in a line within seven feet of the pattern on the wall. This was beyond what the husband was saying about the incident. I had never done this technique before and was concerned about the reliability of the results. I had to consider any other firearm discharge from some earlier date and that some powder may have been moved by medical personnel walking in the room. In my mind, I thought the possibility of transfer was remote, based on the depth of the carpet, but nonetheless I needed to be careful.

By this time, the husband had been interviewed again. He denied any intention to kill her and continued to state that the gun went off in a struggle while they were standing.

Days later I received the rest of the wife's garments for examination. One item was her dark blue bath robe. Because of the color, it was nearly impossible to see any bloodspatter on the garment by visual exam. Besides the close examination of the outer layer, I also examined the inside, which was much lighter in color. I am not sure why, but I reversed her sleeves all the way to the cuffs. To my surprise, there was a small lump of something attached to the inside of the right sleeve. The location would be near the elbow on the underside. The size of this material was three to four millimeters and had an amorphous shape. It was not bloodspatter, but it did look like tissue. Not sure what to do now, I photographed it and placed the garment back in the paper bag and sealed it.

I had a good medical background and was confident that this was tissue, but because it was so small, the origin of the tissue could not be determined visually. I called my friend Dr. Wilson, who had conducted the autopsy and asked him about doing a frozen section on this material. A frozen section is a process where a small amount of tissue, usually from a surgical procedure, is quickly frozen in a special instrument. The frozen tissue is sliced very thin and the slices are placed on a microscope slide for examination under a microscope. This was a common procedure in hospitals.

He could not think of a reason why it would not work, although we were again headed down a path neither of us had traveled. At least to his knowledge, no one had used frozen sections to identify spattered tissue in a criminal case. The next day I took the garment to the hospital; we removed the tissue and Dr. Wilson did a frozen section. The results came back as skin and brain tissue. Wow!

I went back to the lab and discussed the case with my boss, Chuck. We reviewed all the evidence from the beginning. The blood pattern on the wall and the bed put the wife on the floor at the time of the shot. The shot was a long angle from the top of the nose to the top of the head, something certainly possible if the husband fired the shot from a distance beyond her feet. Considering the muzzle at about seven feet, the reach of the arm put the husband at about ten feet if he did it. I knew that early on the bullet had struck the ring on the underside of the left hand. That would place her left hand in the bullet's path, either over the barrel or in a defensive position. A defensive position was looking good at this moment. Then we had the bath robe. When we placed the robe on a mannequin, the sleeves on the bathrobe were oversize and hung below the arm. With the presence of brain tissue, the only way the tissue could get so far up the sleeve was if the sleeve opening was pointed at the brain. The spattered tissue would travel in straight line in such a short distance. This put the right hand close to her right ear, a potential defensive position shown in Figure 3.

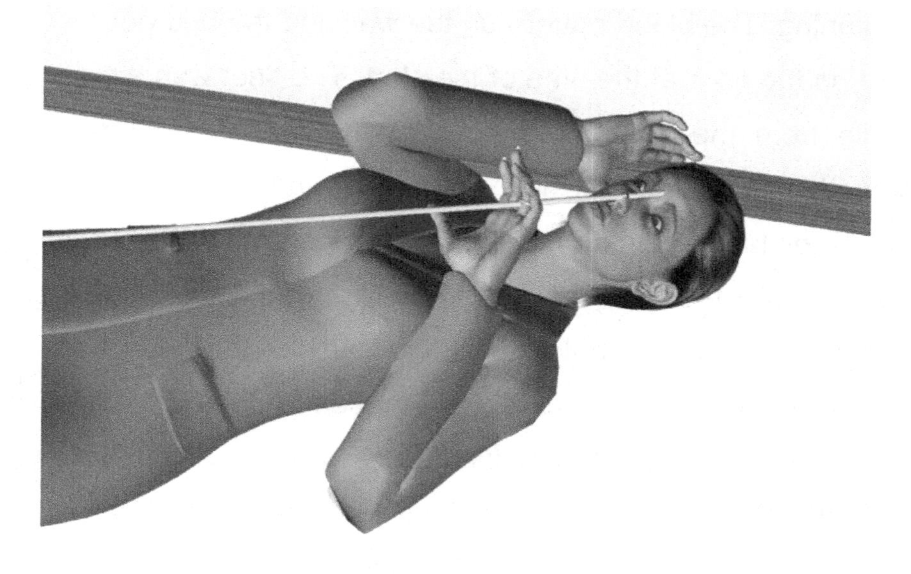

Figure 3

In summary, the combined evidence did not support the description of the incident by the husband. Based on our forensic findings, the husband was arrested for murder. And while we waited for trial, the bedroom evidence remained in the courthouse. I went back and visited it several times because every time the DA had a question, we would go down to the room and review the actual evidence. The plan was to transport our bedroom collection into the courtroom and set it up again at trial.

The husband had a good lawyer, but considering all the evidence we had against him, he decided to plead guilty at the last minute. I was both relieved and excited about the way the case had developed and the outcome, from attempted suicide to a murder conviction. All from a three-month-old scene.

As time went by, there would be many more cases, but this one stuck in my mind as a first, my first on-my-own, forensic investigation.

In later years when I taught crime scene investigation, I would state what I called the Pex Rule:

If it is a witnessed death, you treat it as a homicide until you can prove that it was not.

By doing so, the homicide team would be activated, and a thorough investigation would follow.

Chapter 5 The Elimination Approach to Thinking Like a Forensic Scientist

In the beginning of your training as a forensic scientist, you follow in the footsteps of your mentor and do as instructed. There was no thinking involved. He or she tells you what to pick up, and you do it. Procedure manuals teach you the steps to follow. At some point you will be faced with an unknown situation that was not covered in the procedure manual. In the criminal justice system, what you do has consequences both to you on the witness stand and to the accused. This chapter sets the stage for a method of how to proceed knowing that the Constitution states that a person is innocent until proven guilty. Does this proof come from a single statement of inclusion from your work, or is it by eliminating other potential sources until only one inclusion remains?

In developing a process in which to proceed, were there words spoken a long time ago that still ring true today? In *The Sign of the Four* by Arthur Conan Doyle, Sherlock Holmes says, "How often have I said to you that when you have eliminated the impossible, whatever remains, however improbable, must be the truth?" Is there truth in a fictional story, or did the writer rely on true scientific theory to create the fiction?

At some moment in time, a scientist will reach the point that makes him or her different from the rest of the law enforcement investigative team. What is that difference? It should be how he or she thinks.

In law enforcement, there are often more contacts to be made than there are personnel to give the case what it deserves. As a result, you learn from the first witness what happened at a scene, and then you go looking for evidence or statements to support what happened. That is, until you find that the first witness was wrong. This was deductive reasoning, the process in which a theory was developed, and you seek information to support that theory.

For example, you, the officer, are driving down the street, and over the radio, you hear that there was a bank robbery. The suspect was wearing blue jeans and a red sweat shirt. As you approach the scene, you see someone who matches that description, and you stop him to obtain more information. It would be impractical to drive around, see how many people in the area are wearing the same clothing, and then eliminate those who probably were not involved through other means.

Deductive reasoning is effective and efficient. In the investigation of a crime, officers ultimately ask what, where, when, and by whom as quickly as possible. Their intuition based on training and experience are their guide. Those witnesses who do not provide information supporting the direction of your investigation are set aside, at least initially. As the investigation continues, there is a methodical process, which is discussed later.

As scientists, we learned in college about the scientific method and the steps involved. This method is based on inductive reasoning. The scientist evaluates all of the evidence before he or she reaches a conclusion. This does not remove nonconfirmatory evidence. As it takes longer to process, reports of findings take longer to produce. If we used this method to find a bank robber when the scene was fresh, he would be in the next state by the time everyone else was eliminated.

When a scientist is at a scene and examines the various forms of evidence, such as bloodspatter and bullet holes, he or she should ask several questions in preparation for reconstruction of the events:

1. What are the competing theories?
2. Are the variables controllable?
3. What facts are indisputable?
4. What are facts and what are beliefs of others?

5. Is there a process to resolve the issues?

With the scientific method, we have a process for problem solving. The process looks like this and addresses step five:

1. Make an observation.
2. State the problem to be solved.
3. Develop a preliminary hypothesis.
4. Observe and experiment.
5. Interpret the data.
6. Discussion of results.
7. Make a final conclusion: Is the preliminary hypothesis confirmed, rebutted or neither one?

These steps are general and moving from the words to an action plan can be challenging without some minor changes to meet our specific discipline.

In my business, I offer the same steps with a slight modification. These changes come from a paper written in 1948. Mapheus Smith published "The Method of Elimination in Scientific Study" in the journal *Philosophy of Science*, published by the University of Chicago Press. Smith talks about two approaches of getting to the truth. In the first, he calls it the most adequate idea the mind can perceive without any further examination, sort of the direct act based on your training and education at the time.

This is a long sentence that says that you already have the information and you act on what you know immediately. The other method requires not only to approach the situation based on facts at hand through current knowledge but to also perform tests on all other possibilities to arrive at the truth. This is the search for the truth by elimination. With that thought in mind, one performs tests until no further items are testable. All results that cannot be eliminated constitute the truth, until new information requires additional testing.

The following quotation is from Smith's article. Smith says, *"And it is the logical basis of the legal theory that an accused person is guilty until proved innocent or innocent until proved guilty, since, in theory, the proof in such cases must be inescapable, and can be obtained only by a process of elimination of all other possible explanations."*

The statement should have been written for our profession. However, I realize that the author apparently was never in the courtroom. The "proof" of guilt or innocence is not always inescapable. It is simply the best twelve people can do with the facts at hand at that given moment.

Smith softens his statement later and goes on to say, *"The net result of the method of elimination thus is demonstration of which is the most adequate interpretation. Such a conclusion is inescapable, because all other possibilities of interpretation have been demonstrated to be less adequate."* Some might argue that what is the most adequate interpretation is a function of the jury, not the expert opinion of the witness. Therefore, all theories that cannot be eliminated should be presented despite the initial appearance of being less adequate to the examiner.

Going back to the scientific method with these statements in mind, we see that the following steps become essential, especially at a crime scene. This is my personal approach to problem solving:

1. Make an observation.
2. State the problem: What is the question?
3. Develop a preliminary hypothesis: What do we know now from statements or current knowledge?
4. Observation and experimentation: Examine the evidence and conduct tests.
5. Interpret the data: Study the results. How do they relate? What else needs to be done? What theories can be eliminated? Repeat steps three to five as needed as new questions arise.
6. Discuss: Undergo peer review by someone technically competent.
7. Have a final conclusion: Provide opinion, continue to test your opinion should new information become available.

8. Those theories that cannot be eliminated remain possible. The truth is providing all of the possible answers to the question.

In his article, Smith confirms this with a few final words applicable to our profession: *"It is always permissible, and should be mandatory, to ask, 'What about the other possibilities? Have you considered them all?'"*

Whether you are a scientist or a non-scientist in law enforcement, these concepts above give the best opportunity at arriving at the truth. The side benefit is that it is also excellent preparation for cross-examination in trial. I have had instances on the stand when I could hardly wait for the cross-examining attorney to finish his or her question because I had already considered the question during the work-up and knew the answer. Maybe Arthur Doyle was onto something so many years ago.

In recent years, there has been much discussion about bias in both lay and expert testimony. Bias or some prejudice is found in all of us; accept its presence and move to the next step. The better you can explain the process you used to arrive at your conclusion, the less bias is implied. The fact that you did consider other possibilities and eliminated them by a specific process is the key to reducing the perception of bias in your testimony. Confirmation through peer review is the second step in reducing bias.

Smith talks about bias in the following quote. He says, *"Another important advantage of the method of elimination is the fact that trial of the various possibilities without bias, prejudice or wishful thinking results in objective and impersonal conclusions. In contrast, the use of the method of intuition alone fails to eliminate bias and prejudice, since the intuitive person substantially lacks the ability to criticize himself and his ideas. He cannot recognize his own bias or prejudice."*

The seven steps I listed in the scientific method for forensic scientists rely on the ability to eliminate bias through diligent adherence to methodology and peer review. How many times have you heard an investigator say, "My gut tells me this guy did it"? His or her gut is intuition, and the conclusion may be prejudiced. I saved my butt on several occasions by making the following statement at the end of my reports: "The opinions expressed in this report are subject to change, should new information become available."

To bring home the point, I was approached by an attorney doing a postconviction relief case. Years ago, I had examined the weapon, wounds, and the scene in his case. In my conclusion, I listed that the muzzle of the weapon had to be at least five feet from the victim at the time of discharge based on the absence of gunpowder particulates on the skin. Years passed before I was contacted again and asked if I had seen certain reports before I had formed that conclusion. I had not.

This was a black powder revolver case, and the wad had been recovered seven inches inside the wound.

Black powder weapons require placing the powder in the cylinders manually and placing a piece of cloth between the powder and the ball. This cloth was called a wad. Originally, I had spoken with the medical examiner about the injury, but he did not mention the depth of the wad inside the body in his oral statements. The new information allowed me to change my report without consequence. I conceded that the shot may have been much closer than I stated in the earlier report. Contact or near contact were now choices that could not be eliminated. I could demonstrate in the process where the knowledge from the first study could allow the original conclusion at the time. The new information is acceptable to the process in step seven. I was right both times based on the information available.

There are additional scientific proofs that can be applied to our work, but elimination theory in my opinion is the best approach. When asked on the witness stand, "How did you arrive at your conclusion?" simply state, "I used the scientific method." But be prepared to advise the jury about how you applied the data in a methodical dissertation.

For several years, I co-instructed a course in criminal investigation and also taught a class in criminalistics at the local junior college. Getting students to understand the difference in thinking between police officers and scientists was a constant struggle. But it was more important than memorizing the chapters in the textbook.

It was too easy to give a test and have the students regurgitate what to do in a situation according to the book. I wanted them to consider who else could have done the crime and why they should keep looking. In other words, what process did you use to reach your conclusion? Not only define the process but apply it as well.

During my work as a defense expert for the US Army, at one time I was tasked with doing a scientific study on hand grenades. I am not an expert on hand grenades, and I didn't know of anyone who would specialize in what was needed for answering the question in the case. A few hand grenades had been thrown during an attack at a barracks set up in one of Saddam Hussein's palaces in Iraq. After the attack on the American troops, a search of the grounds revealed the position of the grenade handles, called *spoons*.

Figure 4

A suspect was identified and arrested. Now I was involved in his defense. In Figure 4 is a photo of an M67 hand grenade with the handle (spoon) attached. I was assigned an explosives expert to deal with the grenades, but I had to design and carry out the testing. Would my seven-step model work here, when I had no current knowledge of hand grenades?

The question was, when a grenade was thrown, where do the handles go? There were other questions, too, but let's take them one at a time. Based on knowledge of the scene, there were limiting parameters for the experimental design, such as the distance a person could throw a grenade, physical obstacles that would impair a throw, and obvious areas where a guard would have seen the intruder. Since we did not know how a person may have held the grenade, we would need to address that variable first as shown in Figure 5. So we started out asking the following questions as listed below and then repeated with different positions, left or righthanded, overhand and underhand. We all know the nature of hand grenades.

There were some obvious minimum distances that we need not concern ourselves with, based on the layout of the scene and the nature of the device.

- Experiment #1 Overhand toss with proper hand position.
- Experiment #2 Right hand underhand
- Experiment #3 Overhand fingers on spoon
- Experiment #4 Tower toss -12 feet above ground

Figure 5

In this study, we reviewed what we knew about the scene and army data on grenades to design the throws. A line was made between the explosives expert and a target. The location of each detached spoon was marked along this line for each method of throwing the grenades as seen in Figure 6.

Figure 6

All of the appropriate parameters were measured, such as the distance a soldier could toss the grenade, wind direction, and slope of the ground. Once the testing was complete with smoke grenades, we moved up to fragmentation grenades to confirm our data. If you were in the military and around grenades, you already know the lethality of these devices. The sound and the crater formed in the ground was impressive. I was fortunate in that my video camera, which was set up on a tripod in the open to document the experimentation, never got hit. There is that old saying that close only counts in horseshoes and hand grenades. I am a real believer in the hand-grenade side.

Finally, we gathered all our data, sat down with the lawyers and the explosives expert, and discussed the work completed thus far. Some new ideas were raised, new questions to be answered, and steps three through five were repeated as we eliminated possibilities. When we were satisfied with the testing and discussion of the data, we agreed on what we could not eliminate, and the results were made ready for trial. In the event that new questions should come up or new information become available, we could repeat steps three through five again. The extent of the expert testimony on a question would be limited by the scope of the experimental perimeters each time tests are conducted. When asked a question on the witness stand, the first thought should be, "did

I address this in my testing?"

If the question is outside the testing, beware that a response may be speculative and stated in that manner.

Again, I am not an expert on hand grenades, but I can still develop and proceed with a model to do testing. The explosives guy was the expert, and he testified to all questions related to the hand grenades. My expertise was in designing a methodical approach to do the testing. No conclusions were presented beyond the scope of the experimental testing (step 4). If questions outside the scope of the testing came up in trial, we could go back to step 4 (page 63) or just state that the question was outside the bounds of the testing. Experts often have the desire to opine anyway and here is where things can go very wrong. Three small words are the most difficult for an expert to vocalize:

I don't Know

Chapter 6 The Skull and Tooth Case

One day while the local police were executing a search at a residence for drugs, they found a small skull. The medical examiner confirmed that the skull was from a Caucasian child. The owner of the old mobile home had been using it to support a candle and claimed that he had no idea where it had come from; he wouldn't even state how long he had possessed it.

I was called to the scene and processed the mobile home but did not find anything else of value. In the lab, I found plants and insect casings inside the cranium, so I called an entomologist at the local university for help. Maybe the bugs and plants might tell us something about the origin of the skull. A professor came down and examined the trace evidence, and his examination of this material showed that the skull originated from an environment close to water. Not much help there; that could have been almost anywhere in western Oregon.

Law enforcement was aware that there were a few young girls missing from the community in the last several years. One young girl was missing from her parent's automobile a few years prior, and it was thought that this may have been an abduction.

It was theorized that this skull might be from that girl. There were some witnesses in the area at the time of the possible abduction, so a lineup was set up for the witnesses to see if they could recognize the owner of the mobile home as having been in the vicinity. They needed one more Caucasian man for the lineup. I stripped off my shirt, messed up my hair, and got in line with the other guys, who were transported over from the jail. I was lucky that no one picked me, but the photos made it back to the lab, where they were hung on the wall and labeled as *Public Enemy Number One*. All in a day's work for a young forensic scientist.

The next day I was at the police station again on other duties, and the agency needed to transport the mobile-home owner to the courthouse for arraignment on drug charges. Word got out to the media about the skull, and there were a bunch of media folks all over both exits from the building. So, we hatched a plan. On cue, several of us ran out the front door and headed for our vehicles, and as we were running, someone shouted a location that was across town. The media dashed for their cars, and in a moment, they were gone. No one was there to see the accused be moved to a vehicle and transported to the courthouse. We laughed for days over this one.

With this case, I learned a lot about skulls and the things you could learn from them. The size indicated it was obviously from a child. The shape of the eye orbitals can tell you much about racial origin, as can the shape of the jawbone.

Also, there were some teeth missing from the skull; that proved to be important as well. As it turned out, there were two young girls missing from the same time period. One body was recovered from the Willamette River absent a head.

During the autopsy of the girl from the river, a tooth was found in the esophagus. A well-known local forensic odonatologist was hired by the DA's office to examine the tooth and compare it to the sockets of the missing teeth in the skull. I watched as he placed the tooth in the skull, and he deemed it a match. Thus, the partially decomposed body from the river was linked to the skull and case solved, or was it? After I watched the rough visual comparison, I suggested that we make a mold of the tooth socket in the skull with a special epoxy and compare the mold with the tooth by using a comparison microscope.

This methodology is common for toolmark comparisons such as scratches made by a screwdriver on a door jam. This technique had never been used for this type of examination, but it would give us a magnified side-by-side image to examine. To me, this showed some questionable areas. But I was not an expert on this, and it was still deemed a match by the odontologist. All was well for a few weeks until the accused started to talk, admitted to killing the girl, and identified where the body was dumped. The dump site turned out to be near a creek, not in the river.

My friend and work associate Linton and I packed our bags and headed for the mountains. The reported dump site was high up in the coastal mountains and well off the nearest road.

The area was timbered and had a heavy ground cover of ferns and other plants. There we did a scattered skeleton recovery and found several human bones but no skull. These surface recoveries are time-consuming and are done similar to an archeological dig. It is expected that the longer a body remains in the outside environment, the more scattered the bones become, due to the action of animals and weather. The vegetation is cleared carefully, and each bone found is measured, photographed, and documented on an X-Y axis of a chart. Over the next several days, the bones of a young child were recovered.

There simply was no one else in Oregon who had the skills to compare a tooth to a skull except our local guy, so we were stuck for the moment with conflicting information in the case. It stalled at that point. A few months later, I was selected to attend an FBI academy class in Quantico, Virginia. I had recently read a book on forensic anthropology written by Dr. Lawrence Angel, probably the foremost forensic anthropologist in the world at that time. I suggested to the DA that perhaps I could show the skull to Dr. Angel while I was back there, since he was at the Smithsonian. The meeting was set up, and the skull and the tooth were sent back east. Imagine how much excitement it would have created as checked baggage. I contemplated taking it just for the fun.

After landing, I traveled to the Smithsonian for the meeting. I was so excited that I could not sleep the night before. At the meeting, I found Dr. Angel to be polite and professional, neatly dressed and with graying hair, definitely not an Indiana Jones. Dr. Angel was very patient with me and spent a lot of time explaining all the physical aspects of the skull as related to the child's genetics. This included the overall shape of the skull, eye-socket shape, the rocker jaw, and other features. Then he looked at the tooth, dropped it in each of the tooth sockets, and quickly concluded that it was a nonmatch. There was no doubt in his mind. I was amazed at how quickly he reached his conclusion.

While we were together, we seemed to hit it off well and talked about crime scenes, bones, and their importance to solving crime for what seemed like hours. Then he said, "We have Lucy here. Would you like to see her?" Are you kidding me? One of the most important archeological finds in modern history, you bet! We went in a back room, and he opened a drawer. There she was, the remains of a three-million-year-old person. What a special moment!

I had read much about the discovery of Lucy and her implications for human ancestry.

I will never forget it. It seemed like only a few years ago I was looking at red blood cells under a microscope in a college lab class wishing that I could find more fulfillment in my scientific daydreams. Now in front of me was the sight of a lifetime.

As I stared at the old bones, I had to wonder how her life might have been and how she had died. It brought to mind that when your clan is running from a leopard or a lion, it really stinks to be last.

When I returned to the lab, I talked with the odontologist. He was initially devastated. But mistakes happen, and you move on. At least that is the theory, but it's easier said than done. The disgruntled odontologist was adamant that he was right, but no one was listening. Silence can be the most telling sound that you screwed up. A few years later, he gave a presentation on this case at an American Academy of Forensic Sciences meeting. In the lecture, he used my epoxy casts to show why he thought he was right. It was a good thing there were no other toolmark comparison experts in the room. They might not have agreed with him. During the meeting, I saw him explaining his conclusion to his colleagues, who seemed interested, but I never heard if they offered support.

What is curious about this case is that everyone involved in the prosecution agreed that the accused could have easily been convicted with the information from the first conclusion, despite the fact that it would have been about the wrong victim. What a scary thought. At the time it was an eyeopener for me on the decision-making process for experts and the impact of what we do regarding other people.

Reportedly the accused had kidnapped the child from a vehicle that was parked outside a diner when the parents went inside. The owner of the mobile home eventually pled guilty to murder.

Chapter 7 The Hazards of Drug Labs

In the 1980s, drug labs were rampant throughout Oregon. California had passed laws against ordering and possessing precursor chemicals used to make methamphetamine, so the cooks moved to Oregon. These were operations run by non-chemists handling dangerous chemicals. To make matters worse, they often set up booby traps to protect themselves from other dopers who might want to rob them of their money or product. It was our job to go in with the SWAT people whenever they did a raid on a potential drug lab site to search for the booby traps and control any chemical reactions taking place. It was not uncommon for the meth cooks to take the labels off the chemicals they were using, which left us with a real problem. We are talking about strong acids, combustibles, explosives, and poisonous substances. A routine day at the local drug lab.

In the 1980s, the common method of methamphetamine manufacture used phenyl-2-propanone (P-2-P) and methylamine. P-2-P and methylamine had a very strong odor. In the presence of other chemicals, a reaction took place, and a base form of methamphetamine was produced.

A strong acid was added to precipitate out the drug. Then the slurry was poured through a coffee filter to isolate the final product. It was a yellowish color and contained a large number of other compounds besides meth. It was hard to believe, but this product was ready to go for some cooks. Others would do a cleanup with a solvent until it was at least looking white. All the common solvents are flammable, especially ether. As a vapor, ether was heavier than air and will travel along the floor until it hits an ignition source, such as a pilot light on the hot water heater. Then the whole place blows up. Cooks also collected chemicals not used in manufacturing just to make themselves look knowledgeable to their fellow dopers.

At a scene, you had to rely on your basic knowledge, your senses, and the limited protective gear you had. In the early days, that was normally a set of coveralls and a pair of gloves and eye protection. As soon as the SWAT team made entry, we went in to assess the lab. If the reaction was going, we had to stay there and continue the process until we had a finished product to demonstrate what was going on. This was seldom a safe place to stay, as it was not uncommon to have volatile materials around, such as a leaking fifty-gallon drums of ether. Theoretically one of these containers could easily blow up an entire building just by us turning on the lights. So we used flashlights until the place was well ventilated.

We would stay to finish the reaction, collect, and document the product, and then we would collect all the chemicals for disposal. What do you tell the disposal company when there are no labels? Good luck!

When I got home from these adventures, my clothing often smelled so bad that my wife made me undress in the garage and leave my clothes out there. Some of this clothing we just threw away. But I still needed time to overcome the dizziness and burning eyes. We had no Hazmat gear for use at the scene.

I recall an interesting case near Triangle Lake, Oregon. The cook who often frequented the U of O chemistry department, had a lot of money and had an active interest in lasers. He would talk with professors about the lasers, and then buy expensive equipment for his own experimentation. No one knew the source of his money.

An informant advised the narcotics team, and a search warrant was obtained. A group of us met at the patrol office in preparation. This would include the SWAT team, the fire department, EMTs, chemical-disposal people, and us. The moment arrived for the search, and all of us headed for

Triangle Lake together. There was a two-lane road into the country for about fifty miles. We had at least ten vehicles traveling together. A grade school child could have determined that the cops were coming. When we arrived, the cook and his family were gone. The kitchen stove was hot, so we were not far behind. The lab was in an old building, and we could see the wires from the telephone pole going directly to the door. He had set up a booby trap of 440 volts to the door!

We got the power turned off and finally got into the lab. There were chemicals everywhere, most unrelated to the manufacture of meth. There was a sophisticated laser setup also, but we did not touch it. The chemicals necessary for making meth were present, and we found meth residue on several items. Evidence was seized, and we returned to our lab.

Days later there was a tip that the cook was back at his residence, so we took off again in a slightly smaller procession. But when we got there, he was nowhere to be found. The residence was in a wooded area with a plain view from a neighboring hillside. We quickly became aware that we were sitting ducks from the hillside if shooting started, so we ducked behind the buildings to get out of sight. We needed to search the premises for the cook, so we drew weapons and began the search. It was reasonably uneventful until I opened a door to a small outbuilding, and there stood a scarecrow in men's clothing. I'm not sure how effective the scarecrow was for crows, but it scared the wits out of me. I had pressure on the trigger before I realized that I was about to kill a scarecrow.

I could just imagine how many laughs that would get back at the patrol office if I had fired. But I was not alone; over coffee later others admitted they had encountered the scarecrow with the same response. The resident cook was apprehended a few days later. This individual made bail and became the topic of several newspaper editorials.

His philanthropy with the U of O created a Robin Hood characterization that was hard for us in law enforcement to stomach. In the articles, there was no mention of the havoc that the drug he manufactured had on the community. With trial coming up in the near future, he decided that he would leave the area with his exit popularized in the media. He left, and I believe he was never apprehended again.

Not all of the cooks were amateur chemists. We occasionally encountered organic chemists with PhDs who had just got tired of the poor wages and decided to make some bucks cooking meth for a motorcycle group. Their downfall was usually getting hooked on the product and becoming weird. In the long term, I was told that permanent abnormal mental changes ran high among these people.

In one case I can recall, believing he was going to be attacked, the chemist accumulated a million rounds of ammunition and multiple firearms in his paranoia. In his travels, he picked up a girlfriend and brought her home, much to the consternation of his wife. One night he was shooting at the wife as she ran from room to room in their house, but fortunately for the wife, he missed. He eventually put her in a car, took her to a remote location, and dropped her off. Somewhere in his upbringing, he should have heard the story about women scorned and the things they will do, like call 911 and turn him in. If she had not reported him, I am not sure that law enforcement would have ever picked up on him.

One afternoon I had just returned from court and was in a sport coat and tie when we were advised that there was a wreck on Interstate 5 and that chemicals were involved. I was the only one available, so I went to the scene alone. When I arrived, the fire department was decked out in hazmat gear complete with a self-contained breathing apparatus. We did not have any protective clothing other than a pair of overalls, and they were in another vehicle. A photo of me in my court attire at the scene surrounded by hazmat folks made the front page of the newspaper the next day. The rollover was a small sedan containing two people and lots of drug lab chemicals. The two occupants were transporting a drug lab and were probably overcome with fumes and lost control. The driver, who was described as injured and bloody, managed to flag down a passing motorist and left the scene. The passenger was trapped in the vehicle and abandoned. There was certainly no love there. The driver was never seen again.

The fire department extracted the gal, and she was transported to a local hospital for her injuries. She would not talk to the police and requested an attorney. At this scene, the local fire department had no idea what they were dealing with and relied on me for direction. I can be seen in Figure 7 in a sport coat and tie surrounded by folks in hazmat gear. We did not have anything like this gear available.

There were two major problems and several minor ones. I allowed them to spray down an area that looked OK, but the area began to smoke and caught fire. So much for that idea. I had no idea how that happened. The next problem was a one-pound broken bottle of sodium cyanide with the contents spilled onto the ground lying next to an unbroken container of hydrochloric acid (figure8).

I-5 car crash spills drug ingredients

Driver leaves passenger, thumbs ride from scene after car rolls over

Figure 7

I immediately advised the authorities that they divert traffic a few lanes over on each side. Frankly I was not sure what to do, but something had to be done. I circled the area several times, like a coyote checking out a deer carcass. Finally, I had a fireman get me a box filled with kitty litter. I put on a pair of gloves and stepped up to the bottle from upwind. There was no time to waste. I picked up the bottle of muriatic acid and placed it in the box and asked the fireman to remove it until we could decide what to do next. The fire department sprayed down the area with a hose until all sign of the powder was gone.

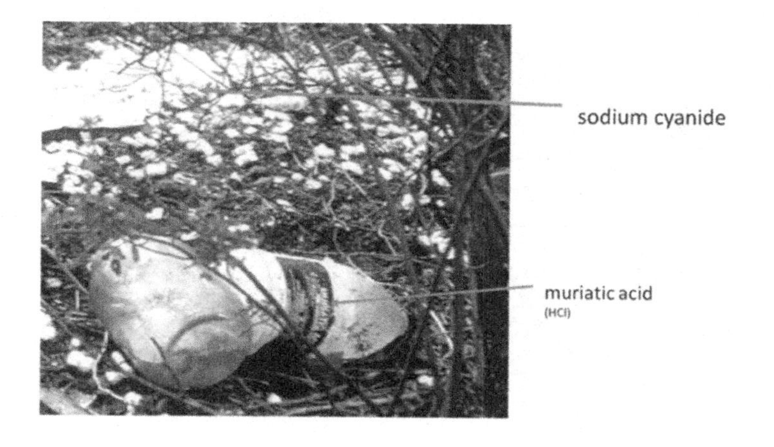

Figure 8

As I reflect on this, my thoughts are, *Stupid! Stupid! Stupid!* With this much cyanide, there was no backup plan and no margin for error. Death could have been the result for several of us at the scene. It was a long time before I ever told anyone about how stupid I was at this scene. But I had no academic or informal training that prepared me for this situation with the amount of poison present.

The next problem was a stick of a smoldering substance lying on the ground. I knew it was one of two possible substances, either elemental lithium or sodium. One is stored in water and the other in kerosene. It can be flammable or explosive if you get it wrong. I sent a fireman to get me a gallon of kerosene on a hunch. We approached the stick, and I flicked a little water onto it. It popped and acted like sodium might, but how can you be sure? I told everyone to stand back, put on a couple layers of gloves, picked up the stick, and placed it in the kerosene. Nothing happened; fortunately, I got it right. Again there was no one else to make these decisions, and I was lucky, damned lucky. In reflection, did I mention that I was stupid? I don't recall any chemistry classes in college that prepared me for this, but something had to be done.

My knowledge came from many previous drug lab scenes and having seen these chemicals before.

Sometimes the good Lord is looking out for you. I walked away from this scene unharmed. Not long after this case was over, the police union took up our cause and got the department to buy us the proper gear for handling a scene like this. They used the photograph from the *Register Guard* to accomplish that task.

In our profession and dealing with drug labs over the years, there was a common observation we were all aware of: There were no old drug lab cooks. Undercover agents tracked these people. They were usually dead within five years of their initial manufacturing crimes. Besides the primary product drug, there were many other chemical compounds created from the clandestine reactions that were ingested as a minor component of the sample sold. There were no studies that determined just how dangerous these compounds were to the human body, but we had a good idea. If you extract a rough sample of methamphetamine and run it through a gas chromatograph, there would be a single peak for meth and lots of other peaks for compounds that were unidentified, and I have no idea how they might affect the user long term. Looking at these charts, it was easy to see why there were no old cooks. For us, it was not uncommon to see our lab guys under the weather for a couple days from responding to an unventilated clandestine lab. But we moved on, the state did not provide us any options.

It is important to note that most of us had a chemistry background and were as cautious as possible when dealing with these labs. Within state government, there was no one else who had our experience and knowledge. Consider the rollover on I-5; no one else saw the sodium cyanide next to the muriatic acid. Our presence not only saved lives in law enforcement but was pivotal in getting successful prosecutions. Some risk was unavoidable.

Chapter 8 Mystery at the Rest Area

Chuck came into the drug analysis area where I was working and indicated we had a scene at a local rest area. This was a major public rest area adjacent to Interstate 5. A person had called into the OSP that one of the stalls in the lady's restroom was covered with blood. A trooper had gone in and confirmed the call; we definitely had a scene. A sample of the blood was being transported in for us to examine for species origin. When the blood was logged into the lab, I took a portion and ran the species analysis. It was human blood. No other information was available.

My associate Linton and I were charged with processing the scene, knowing that this was a potential homicide scene. It was suspected that either a serious altercation had occurred in the stall or that body parts may have been flushed down the toilet. The scene was secured by OSP when we got there. When we looked in the stall, there was blood on the toilet and the surrounding walls and initially what we thought was bloodspatter on the ceiling. Bloody hand swipes were visible on the walls as well. First, we had a fingerprint tech print the stall inside and outside. None of the prints developed were in blood and were held for future reference if needed.

As it was a public restroom, the value of the prints was unknown, and no recent time frame could be assigned.

Linton and I photographed the entire scene, collected more blood, and took copious notes. When I took a closer look at the spatter on the ceiling, it was apparent that this was not spatter from an assault. There were numerous flies in the restroom that were landing on the blood and then flying to the ceiling. The flies created the small dots of blood seen on the ceiling, it was not a bloodspatter pattern. The characteristic nature of these dots was that they never exhibited any directionality and were always perfectly circular throughout the pattern.

What was missing at this scene were any potential body parts, weapons, clothing, or even a blood trail leading into the stall. But could it have been a baby? If it was a newborn, I had a test for that. I went back to the lab and ran a fetal hemoglobin test on the spectrophotometer, but it came back negative. There was no sense in running genetic markers yet, as I had no one for comparison to the blood. All of us at the scene held an impromptu meeting on what we had. The last question to be answered was, what was in the huge holding tank besides poop? And how were we going to find out?

To start, we thought we could drain the tank, spread the contents out somewhere, and see what was in it.

We talked to a septic disposal company, and they agreed to drain the huge tank. A few hours later, several trucks showed up on scene and began pumping.

The trucks went to a local county dump, where they lined up and, one by one, drove slowly around in a circle depositing their contents while we walked along behind and looked for body parts. I don't think I ever wanted to know what was in a rest area septic tank and I was telling myself that I still did not want to know as I walked along. This was not what I had signed up for when I had joined the OSP, but this work just came with the job. By the time the last truck emptied, I had seen a few blood clots but nothing else. This was a real enigma. There was too much blood lost to be a cut lip or a bloody nose, and the possibilities included homicide, assault, or a prank. The media had picked up on the case, and television crews were covering it with an OSP spokesperson providing details. I can just imagine the questions: "you looked in the septic tank and what did you find?"

One of the detectives took a flashlight and peeked into the tank; it was not empty. Now what to do? Linton looked at me and said it was our responsibility to go into the tank. I told Linton that if he wanted to volunteer, go ahead but to leave me out of it. I suggested we just take a wait and see attitude and see what developed. The bottom of the tank was still knee deep in poop and a million other personal items that I don't care to describe.

After the detectives talked it over, they decided that two of them would go down inside. There was a concern for methane gas, so SCBA gear was required. I looked at Linton and said, "Stay tuned; this should be really interesting." One of the designated detectives was my friend Dennis. Dennis was a funny guy and always cracking jokes. I have a weak stomach, but mine was nothing compared to Dennis's. He and I had gone fishing a few weeks prior, and I had to clean his fish! I had thought he was kidding when I told him we need to clean the fish, and he said he couldn't; his wife always cleaned the fish. I browbeat him into trying, and he got the dry heaves when the guts were exposed. It was unbelievable. I finally took over and cleaned the fish.

Now Dennis was suiting up in SCBA gear with air tanks, a full zip-up white coverall, and hip boots. He laughed and joked all the time he was gearing up. The other detective was laughing, too, both talking about this being about the weirdest thing they had ever done. The joking finally stopped when the airtight helmet went on, and they picked up an aluminum ladder. They pushed the ladder down into the muck, ran an extension cord with a light down the entrance hole, and started down the ladder. Mike, the other detective, was down the ladder in an instant, but Dennis was still circling the hole in the top of the tank. Finally, he gave everyone a big salute and started down and eventually disappeared from sight.

I looked at Linton and the others, and we were all silent with anticipation. In less than a minute, Dennis came roaring up the ladder, removed the helmet, and puked. We all laughed as he lost his breakfast, lunch, and probably the dinner from the previous evening. He heaved over and over for several minutes before regaining his composure. At this point he put the helmet back on and headed back for the ladder. Mike was still down there. Dennis gave the usual wave, smiled, and descended into the hole. As soon as he disappeared, we laughed ourselves silly.

As expected, in less than a minute, he was back up the ladder, ripping off the helmet and puking again, much to our entertainment. This time when he came up the ladder, he hooked the extension cord with his foot and broke the light. Mike was still down there in the dark, knee-deep in poop. It quickly ran through my mind that they might think they needed a replacement for Dennis, and I did not want that to be me. I quickly suggested that we rescue Mike, get him back in fresh air, and discuss this thing. What were we going to do with the stuff still in the tank? Push it around by hand? From a biohazard standpoint, it did not get much worse than this.

We shined a flashlight into the hole, and Mike found his way back up the ladder. Dennis was beside himself, still cracking jokes, but it was clear he was not going back. That was too bad; I had not laughed so hard in ages.

We decided that we needed more supporting information that body parts might be in there before we would continue. Eventually the search was discontinued, and the rest area was reopened to the public after the mess was cleaned up.

That evening the story about a potential homicide at a rest area made the news. Late that evening the patrol office got a call from a concerned citizen who lived on the coast.

His wife had had a tonsillectomy at a Eugene hospital, and when they were returning to the coast, she had started bleeding and asked to stop at the rest area. The woman had swallowed a fair amount of blood before she entered the stall in the restroom and began to vomit. She had felt very light-headed. She said she vomited several times and had blood on her hands as well. She stated she had difficulty supporting herself as she vomited and feared that she may have missed the toilet. No kidding! After several purges, she went back to her car, and her husband drove her home. They were sorry, but she did not feel well enough to clean up the blood in the stall.

We were quite relieved; at least the mystery was solved, and we did not have to go back to search the rest of the septic tank. I teased Dennis for years that not everyone gets to go spelunking at the local rest area.

Despite the early experience into deep doo-doo, Dennis was promoted years later to deputy superintendent of the state police, and after that, he was named by the governor to be director of the state prison system. The prison system was embroiled in controversy when he took over, and there he was, knee-deep in poop again.

Chapter 9 Body-Fluid Testing in the Years Before DNA

Because of my background in the medical field, expanding the crime lab role in genetic testing was a primary responsibility for me. As was often said, this is what they hired me for. Cases often hung in the balance based on the ABO type or other genetic markers. This was long before DNA testing was a whisper in a geneticist's ear. This chapter chronicles the state of the biological testing in the late 1970s and 1980s Biological fluids associated with crimes included saliva, semen, vaginal fluids, and blood. Our methods were not sensitive enough for sweat. Discussing these tests and their applications in the early days helps describe the amount of work required in those days compared to modern DNA testing.

Some body fluid testing was done by everyone in the crime labs, despite limited knowledge of biological testing by some of the folks. The attitude of management was that everyone should be capable of doing most tests. Each person was in the business of solving crime. Involving others outside the state was cost prohibitive. For example, the testing of dried blood required a visual test, a color presumptive test, and confirmation by determining if it was of human origin.

The presumptive color test was done with a reagent that contained benzene, a known carcinogen in daily use by all of us. There was also a crystal test for hemoglobin using a chemical called pyridine. If the amount of blood was great enough, one could dilute it in water, run a spectrum in a spectrophotometer, and get the characteristic peak around 540 nanometers for oxyhemoglobin. The human origin test was done by immunodiffusion in a petri dish with an antihuman antibody generated by injecting human blood into a rabbit. The rabbit would eventually be bled, and the rabbit's blood would be extracted, processed, and used for this test. The immunodiffusion test took overnight to complete. This step was usually done on a couple bloodstains, and the rest were not tested. Since this test was slow and may have to be repeated if it did not work, no results went out until this was complete. Everyone in the lab could run this test but results might take a couple of days. Meanwhile, investigations might be held up.

But when I introduced a new method of human origin testing called *antihuman globulin inhibition testing*, I was on my own. The test took about thirty minutes compared to overnight and was more sensitive than the immunodiffusion method, but it was a little more complex to run and to explain in court. Consequently, I would end up running the test for everyone else in the lab if my coworkers were in a hurry.

With this test, I could run several stains in short order. The speed of an investigation just got a boost. After one determined that the bloodstain was human in origin, the next step was to attempt an ABO type. For this, one needed indicator cells in the four common types in the ABO system, which I procured from the local hospital lab. The oldest and simplest method of testing was called the *Lattes Crust test*. The test relied on naturally occurring antibodies found in blood plasma. If your blood type is A, you carry anti-B in your plasma. Type B carries anti-A. People who are O have both, and type AB have no A or B in the blood plasma. The test was fairly simple, in that blood crust was scraped from a stain and placed on a microscope slide and covered with a cover slip. This was done on three slides. On each slide a dilute concentration of type A, B, and O red blood cells was added. Then one waited to see if serum leaked from the crust and caused individual red cells to *agglutinate*, a big word for clumping together.

If one of the slides did, one could determine type A or B. If both slides showed agglutination, the type was O. If nothing happened, in theory, the stain could be type AB, or the test did not work. If you knew that one of the victims in a crime was type AB and you did not get a result, your report reflected just that; either the person was type AB, or no result was obtained. With only this test, we frequently did not get a reliable result of any type. The sometimes-poor quality of bloodstains in a homicide case was often an issue, making final results questionable. Remember this when we get to the Reavley case later in the book.

The department was anxious to improve testing in this area, and this is what they hoped to get from me. The first step was to educate the people doing the testing. The Lattes Crust test had reliability problems and was not very sensitive. So I worked up a capillary-tube reverse-typing method that was far more sensitive and easier to interpret. Again, I could teach it, but usually I had to run it too, in casework. After the first few years in the lab, I was slowly becoming a forensic serologist, one who specializes in biological testing.

The primitive testing available in forensic serology became even more evident when I was receiving training in sexual assault cases. Looking for semen on the victims' evidence, such as panties, was primary to case resolution. In my first day of training, a pair of panties was examined for semen in the area between the leg openings. We had a reliable test for the liquid fraction of semen; it was called the *prostatic acid phosphatase test*. Unfortunately, the result was also positive in the presence of high concentrations of yeast. Did I mention that people with diabetes are prone to yeast infections?

Within the group of professionals, I worked with, I was the only one who knew that fact. Thank goodness for my medical experience. Take a pair of panties and leave them in a sealed plastic evidence bag, which could create a moist environment, and you have a problem. Hospitals were notorious for putting clothing in plastic bags. The solution was simple; look for spermatozoa. I was blown away when my instructor took an extract of a potential semen stain, placed a drop on a slide, and tried to look for spermatozoa by what we call a *wet prep* under a microscope. This was beyond primitive, it was totally unreliable.

If the sperm were alive, this method was OK, as you could see them move. In our case, they were always dead and often looked like other debris recovered from clothing. What was worse was that sometimes debris looked like sperm. It was an OMG moment, and during my lunch hour, I went to a local clinical lab and procured some special stains used for staining human tissues. I had used these stains before and they gave spermatozoa a definitive color, making them easier to see. I went back to the lab and set up a staining procedure so spermatozoa could easily be seen under a microscope if present. Soon the entire lab system was using the staining technique, and forensic science in Oregon moved a tiny step forward that day.

In 1980, I had only been out of the clinical lab for a couple years and had kept in contact with my friends still working in the medical profession. One close friend had left the hospital environment and went to work in a regional blood bank working on special projects dealing with tissue typing.

Over lunch one day, I posed the question with him of doing tissue typing on spermatozoa from a sexual assault and identifying the assailant's genetic markers. Both of us had a good background in immunochemistry and began drawing various procedural pathways to identify the HLA type of a sperm with HLA antigens.

HLA stands for human leukocyte antigen and is a tissue typing system that offers specific markers on white blood cells and other tissues. This system was used to match donated organs to recipient people hoping to avoid tissue rejection when the new liver or other organ was transplanted during surgery. The procedure had to identify the sperm and be sensitive enough to see the reaction of the antibody to antigens on the sperm magnified four hundred times under a microscope. We settled on a procedure called *immunofluorescence*. One did the reactions on a microscope slide holding the sperm, and in the final step, a fluorescent tag was applied where a specific reaction took place. If a sperm was found and was positive for the specific antigen, it would fluoresce, making it readily visible by a specialized microscope. This would be a genetic blood type that could be compared to a donor. Perhaps the first application of this method anywhere in the world for sexual assault examination.

My friend went back to the blood bank and did a few preliminary tests, and they worked. Keep in mind, it took a couple more weeks to get all the bugs out of the procedure.

Now we needed control specimens. That would be a blood sample and a semen sample from male donors. The blood would be typed first to determine the HLA pattern, and then a semen preparation was dried on a microscope slide for the new procedure. We kicked around ideas about where to get some samples and were stumped. It was a little difficult to approach your guy friends and say, "Would you mind masturbating in a test tube for me?" We were stuck. We fretted over this for days, and then I recalled that a physician friend of mine was working at a plasma donor center in Eugene. Maybe he could help.

I called him, and we talked about the specimen problem. He had a great idea. If we offered an extra ten dollars, he was sure that a number of the local plasma donors would comply and provide a semen sample. The blood bank was willing to pay, and suddenly we were getting a number of blood and semen specimens every week. I would pick them up in Eugene and ship them to the blood bank in Portland. By now the blood bank management was involved, and they authorized a technician. She was trained to do all the testing so that my friend and I could focus on the results. I could keep up with my case responsibilities while the blood bank did all the work.

In the next few weeks, the testing was going extremely well. I would call up the same day that I sent the samples, and they would already have results. The testing was working so well, I could hardly believe it.

Everything was matching up, so it was time to do some post-coitus testing. There in was another dilemma: “Pardon me, ma’am, but have you had sex recently? Perhaps you could provide me with a vaginal swab?” We fretted some more, but after much wringing of hands, my friend put the word out at the blood bank. I did the same in the crime labs. To my surprise, we got a number of vaginal swabs and blood samples from the women. There was a lot of support for the testing among the staff, everyone was excited and optimistic.

My friend and I would just shake our heads. How did we ever get ourselves in a position to be asking for vaginal swabs from women after they had intercourse? I had to ask myself how I always ended up in these weird predicaments, and then I reflected back on my college days looking for cockroaches. My old friend from college, Dale, would not have been surprised if he knew. These requests also caused a lot of giggles and jokes in the lab, which I had to endure.

This was supposed to be about immunofluorescence; now it was about sex. I was visualizing a cartoon with me hanging out at a local bar with a swab, syringe, and bottle of whiskey and saying, “Hello, ma’am, like a free drink?” “How’s you love life?”

The moment I will never forget is when I asked for samples from an attractive lady whom I knew pretty well. She said bluntly, “Certainly. I almost never have a semen-free vagina. Would you like the samples now?” I am seldom at a loss for words, but that was one of those moments.

My mind was in overdrive; my mouth was stuck in neutral. In the years to follow, that was the first thing that popped into my mind whenever I saw her; I could not help it.

Testing was going very well. All the samples were working out properly. The management at the blood bank was talking up the new science, and one day I saw reference to the study in *Time* magazine as the new trend in forensic science. OSP headquarters was also excited, and I was getting all the time I needed to complete this project. The problem was that this was going so well that I was getting suspicious. We had gone from a diagram on a napkin to several hundred samples in six months, and they all worked? That was possible but not probable.

I called my friend and suggested that we take a run at this by ourselves on a weekend just to be sure all was OK. He agreed. I drove to Portland, and we spent the weekend running the tests ourselves. And there it happened. Some of the tests did not work at all. Some of them did, and the HLA type of the spermatozoa on the slides could easily be determined under the microscope when it did work. The promise of success was still alive. When we got to the post-coitus samples, some were positive for everything with unreadable results. We kept quiet about this for a week while we tried to determine the problem. Finally, my friend confronted the technician, and she admitted that she had faked some of the results. He asked for a reason, and all she could say was that she wanted it to work.

This was catastrophic as far as I was concerned. The test definitely had promise, but everything that had been done so far had to be tossed. We were back to square one. We did not know why the test worked some of the times but failed other times. So an intense literature search was done, and we found what was probably the problem. In the human-fertility literature, a paper stated that some women have antisperm antibodies that are immunoglobulin proteins similar to our last step in our procedure, so the fluorescent dye was going to stick to them just as well as it should if the test had worked properly and stuck to an HLA site.

In essence, the dye attached to the antisperm antibody just as well as to the HLA antibody. The woman's own antisperm antibody attached to the donor sperm in vivo before our sample was even collected. We had no way to control this and were momentarily stumped. I reported the false results of testing to my superiors, and despite the potential, they wanted me out of there immediately. My involvement ended, and according to my friend, the entire project was scrapped. I did not have any further official involvement with the blood bank. My friend and I were both very pissed. It was all for not.

But, unknown to me, it was not over for everyone. Many years later, I discovered that the management at the blood bank published a research paper in the *Journal of Human Immunology* and I was a coauthor. What a surprise! I was never contacted about publishing and would not have consented to be a coauthor had I known. My friend who shared in the idea and development had left the blood bank, and he was not listed at all. How unusual. The paper was written in such a manner as to list the successes of the project and not the mishaps. My first publication, and I did not know it existed until recently. In the years to follow, I would write several research articles that were listed on my résumé. I never listed this one.

Years later a similar procedure, called *DQ alpha testing*, was approved for crime labs and did not have the problems we encountered. DQ alpha was a precursor to modern DNA testing.

As a result of my improvements to the lab from my own research, I was getting more and more cases assigned to me, primarily sexual assaults and homicides. The shining light in my successes in methodology was also due to a policy directive from the division director. He gave me four hours a week to set aside casework and do research into new methods. No one else got this opportunity, and my associates supported me as long as I worked their blood-typing cases for them. I would spend the early part of the week ordering reagents and reviewing available literature from home. When the Friday afternoon time became available, I was ready to go.

Since very few of my associates had a background like mine, I would seek scientists at the FBI and in other states as resources. Over the phone, we would discuss the theoretical aspects of blood typing and the problems incurred. As we developed ideas, I was in the best position to put the testing into action. Even the FBI people did not have the time that I had available to put the ideas into action. One of my favorite resources was a British researcher named Brian Wraxall, living in California. We would communicate almost weekly on issues in forensic serology. It was an exciting time.

These advances came to the forefront when we had a serial rapist working in our local university neighborhood. Other than saying it was a non-vasectomized male due to the presence of spermatozoa, we did not have much to offer. However, it was known that about three-quarters of the population secrete their ABO blood type into all of their body fluids, including semen. The fact that one-quarter did not made conclusions more complex. If you got a negative result, was it because the people were non-secretors or because the test did not work?

Semen stains derived from a sexual assault kit containing vaginal swabs were usually a mixture of body fluids from both the victim and the assailant. If you got a type and both people were the same type, again what did you learn?

Nothing was established without knowledge of the secretor status of the people involved. The literature had methods for secretor typing that were single test tube tests. If you got a result or not, you did not know if the people involved were secretors or if the amount of semen was so small that you would not get a result at the concentration of reagents in the test. No one knew. I modified my testing by doing serial dilutions of the reagents. This was much more sensitive and allowed for variability in blood-type concentrations within the evidence material. With a serial dilution, there was a better chance of matching the correct amount of antibody with the antigen.

In addition, the red cells in our body have several antigen sites for blood groups not as commonly recognized as the ABO blood group. One of those antigen sites was the Lewis blood group system. The Lewis system is genetically tied to one's secretor status. So, if you have a blood sample from the suspect and victim, you could determine the Lewis type and potentially determine the secretor status for both parties. Let's suppose that the victim of a sex crime is type A and a nonsecretor and that the accused is also type A and a secretor. If you found type A in your testing of the semen stain, the assailant could not be excluded as the origin of the type A. Without secretor status, the result would be assigned to the victim.

In one of the attacks by the serial rapist, he raided the refrigerator after he committed the crime and took a bite out of a block of cheese. Since saliva also contains the blood type of a secretor, I swabbed the bite mark and did my testing. I got a type A from my testing and tested some of the other cases and also get a type A from them. Not long after the initial testing, a person was caught attempting a sexual assault in the university area, and he was a type A secretor.

An unmentioned highlight of developing this testing was the number of suspects who had their blood drawn and did not meet the blood typing criteria; they were quickly dismissed. Exclusions were absolute. A forensic odontologist examined the bitemark on the cheese and compared it to a mold of the accused teeth. He concluded that the bitemark was made by the accused.

Times were changing. I was much happier with my career change, and the department was also happy with their investment in me. However, constant call-outs for a variety of different crimes was still a problem, but we complied as requested. I felt good about the progress I was making and the assistance I was giving to the criminal-justice system. There were times I would be on a call-out most of the evening, bring the evidence back to the lab, work all night doing blood typing, and have the results ready for the investigators' meeting in the morning. Then I would work for the rest of the day.

One night in the early 1980s, Linton and I were called out to Corvallis; a rape had occurred at a fast-food restaurant. A rape kit was obtained from the victim, and I was able to get a B secreted type for the assailant. There had been several rapes and homicides up and down the I-5 corridor for over a year, and my boss had convened a meeting of detectives for each of the areas to compare notes. Detectives reported that someone had heard the sound of a Volkswagen leaving one of the scenes, another clue. The fact that the assailant put a Band-Aid across his nose was a common element in all the crimes. Similar crimes had occurred in Oregon, Washington, and California. It was apparent that we were dealing with a serial killer or rapist nicknamed the I-5 Bandit. Some of the rape victims did not survive.

In another important clue, I had examined the bullet from one related case in which two women were shot. One of the women had died, but with the other woman, the bullet was such low velocity that it bounced off her skull and did not penetrate. She was knocked unconscious but survived. My testing showed that the all-lead bullet was .32 caliber. Based on all-lead and its diameter, I suspected the weapon was probably a revolver chambered for the .32 S&W long or short cartridge. This was an old weapon not commonly used in recent times and an uncommon cartridge today because of its low performance characteristics.

Weeks later the local detectives got a lead on a guy who happened to live in Eugene, and we rushed over to serve a search warrant on his apartment. His name was Randall Woodfield.

There was an initial standoff with the police, and Woodfield apparently used that time to burn several items in his possession. But when we searched his gym bag, under the bottom flap, we found a single .32 S&W long cartridge in the bottom of the bag. I knew we had the right guy at that moment. Guess what? He was type B and owned a Volkswagen. The revolver was nowhere to be found. Woodfield would make phone calls immediately after committing a crime, and the date and time of the calls were on his phone bill. This put him not only at a location but at the time of the crime.

He was found guilty on several counts of rape and murder in subsequent trials based on our findings and other evidence. Ann Rule wrote a book about this case, and it was titled: *The I-5 Killer*.

The importance of genetic typing was making a mark in Oregon casework, but it made an unusual turn in the next case.

One morning at work, I got a call from one of the labs in eastern Oregon on yet-another interesting case. They were working on a sexual assault involving a fifteen-year-old girl, and the accused was a male twice her age. The girl was pregnant from the abuse, and a question arose on whether we could tie the male to the child. Just one problem: He was vasectomized and claimed that he could not be the father of the child. Normally after surgery, the patient submits a sample for testing to ensure there are no little wigglies still around.

Hospital records confirmed he'd had a vasectomy, but no test results.

The girl submitted to a late-term abortion, and a sample of blood in a test tube was collected from the baby, the mother, and later the potential father via a warrant. The samples were sent to me, and I took them over to the local hospital, where I ran a series of red-cell-antigen tests on all three samples. Since I had worked in a blood bank, I was knowledgeable on the various inherited red-cell antigens; all I needed were the reagents. Most people are familiar with the ABO blood-group system, but actually there are many lesser known blood-group systems. Each one has a specific determinable type based on inheritance. The result of the testing was that the accused had a high potential to be the father and could not be excluded. These tests are never absolute, and the problem still existed about the vasectomy. He was not going to give a semen sample for us to find out, and the DA was beside himself on how to force getting one. But we were going to trial anyway.

There was the normal procession of witnesses, and I was going to be last. Then there was a surprise. Keep in mind, this happened in a small rural community with only a few local physicians. The physician who performed the vasectomy was called to the stand, and under oath, he admitted that he had operated on the accused but that once inside, he could not remember which tubes to cut. So he had stitched him back up, did not tell the man of his incompetence in the surgery, and charged the defendant for the procedure anyway. The courtroom was dead silent as jaws dropped. With this new information, the accused was convicted. My testimony only complemented the physician's statements. Whoever initially made the statement, "*Truth is stranger than fiction*", had to have worked in law enforcement.

In the following years, besides ABO typing, there was a process called *electrophoresis* that also was a useful tool in forensic serology. In graduate school and the clinical lab, I had done several electrophoretic procedures and was well prepared to utilize this procedure in the crime lab.

Now I could do immunoelectophoresis and get results within an hour when other methods may have taken all day. Electrophoresis was the preparation of a gel similar to Jell-O on a glass plate. After I applied proteins at one end, the entire plate was charged with an electrical current. It is actually more complex to run, but that was the basics of what happens. The proteins were charged particles in a mixture such as blood and would separate out into bands of individual proteins if left exposed to the electrical current for some time. The distance the bands migrated aided in the identification of the protein when compared to known standards.

We were at the cutting edge of this methodology when I was in graduate school at San Francisco State, so I was already trained.

The "immuno" side was adding antiserum and having an immune reaction as one of the last steps in the test.

When I reflect back about the interesting times in graduate school, some days it was hard to go home. I was wrapped up in making different concentrations of gels, experimenting with application methods for the proteins, and changing the electrical charge to see what would happen with our selected target proteins. San Francisco State University was also the place for riots in the 1960s. Before I went home, I had to go to the top of the building to see where the riots were happening so that I could find another route off campus.

In the mid-1980s, the crime labs across the country were introduced to enzyme phenotyping by electrophoresis. Enzymes are proteins, too, and the structure of some of these proteins was controlled by genetics, as in blood types. There are hundreds of "blood types" within the human body; some are in high enough concentration to be useful for comparison to a person's blood or other biological fluid. Researchers have studied the frequency of occurrence of a specific type of enzyme in the general population. For example, people who are type A represent 42 percent of the population. This is important, because if you run several different proteins and get a type for each, you can multiply the frequency of each type times the other. For example, if you got type A on a bloodstain that was 42 percent and did another type that was 10 percent in the population, 0.42 X 0.10 = 0.042, or 4.2 percent, of the population would have both types. If you run a half-dozen different tests, the total frequency for that person could easily be below 1 percent.

Low-frequency numbers were never considered a match but simply added to the evidence when presented to the jury.

The FBI offered a two-week course at Quantico in enzyme phenotyping, and I was invited to attend. I knew how to run the tests, but it was an opportunity to hone my skills and hear about any pitfalls the FBI may have encountered. The first week was uneventful. I was assigned a lab partner, a woman from Wisconsin, and we ran the various procedures each day as assigned. For samples, class members had their blood drawn and processed. The methodologies we studied included hemoglobin typing. There are several forms of hemoglobin among humans and this protein carries oxygen from the lungs to the tissues. The most common is type A for adult. There is also a type S found in some people with African heritage and is related to sickle cell anemia, a very serious disease. With the samples we provided, we had all the common hemoglobin types in the lab for testing.

One day when my lab partner, who was very much Caucasian, was out of the lab momentarily, I put a drop of S hemoglobin in her blood standard, knowing that we always ran our own blood as standards when performing these tests. When we finished the first run, I let her interpret the results. I had informed my instructor of my misdeed, and we were both standing back just watching. She was diligent in documenting the results but kept going back to the plate and looking again. We asked if there was a problem, and she said there must have been some contamination or mistake in application of the samples; the one that was supposed to be hers had an S hemoglobin band. The instructor suggested she rerun the samples to determine the source of the error. We turned and grinned at each other as she set up a new plate.

Thirty minutes later the run was complete, and she was looking over the results. Knowing for certain which blood sample was hers, she looked up from the plate and said that she couldn't believe it. You could see the emotion developing in her face, and she suddenly exclaimed, "My momma told me we were part Cherokee!" I quickly turned away and laughed myself half-silly, as did the instructor. I waited a few minutes, but just to be sure she did not make a quick phone call home, I told her I had spiked her blood sample. I am not the best at nonverbal communication, but she looked at me like, "If there is any way I can get you before I walk out of here, I am going to find it." I was looking over my shoulder for the rest of the class. She never got even and in time forgave me for my sin, and we finally had a good laugh. It is just the way I am; I cannot help it.

Getting back to working in the lab, one of the areas for specialization under serology was the examination of sexual assault evidence. When I first started in the lab, we did not have a kit. The evidence received into the lab was dependent on the knowledge of the physician who performed the sexual assault examination. Sometimes we got slides with vaginal content to examine; sometimes we got swabs.

Looking for foreign pubic hairs was not even in the ballpark. What was obtained as evidence was so variable that the success of the prosecution was unfairly based on the quality of the sexual assault exam in the hospital. In some cases, evidence was collected and held in the ER, never to be turned over to the police.

I put together a kit complete with instructions for distribution to the local hospitals, and the quality of evidence received made a big turnaround. This included me going to the local ER's and talking to the staff about proper performance of sexual assault exams. I would have to hit all three shifts in the hospitals to be successful. Since I had worked ten years in hospitals, I knew my way around them and the way to make contacts that would prove useful. During some of these lectures, I would have a chance to talk privately with the staff.

I was absolutely amazed by the number of times I heard from nurses that the assault victims probably had it coming. What can you say?

It was years before I stopped hearing this statement routinely, and hopefully our work in the labs and media exposure of these serious crimes finally made a difference.

By the mid-1980s, I was working serology cases almost exclusively. But despite other advancements made in the lab, there were always problems to be solved. Reflecting back, I often laugh at a terminology problem we encountered. What do you call the area in the middle of a pair of panties or undershorts? The crotch?

The word was derived from *crutch*, a forked stick used for farming. I had a thing about using that word in a professional report, so what do you call it? Was it the area at the bottom of the abdomen or the top of the thighs? We kicked around different words, and I finally settled on," *the area between the leg openings*." That seemed much more professional to me than *crotch*. I could remotely see crotch for an adult, but what about a diaper? Yes, we occasionally looked for semen and or blood in a diaper, sad as that concept sounds. So, "*the area between the leg openings*", became the norm. At times I had to ask myself, "Do any normal people have these conversations?" Then again, in this business, what is normal?

On the instruction form I created, I asked for the time of the incident and time of collection in order to get an idea of the amount of time that passed before the evidence was collected. This helped when you did not find any semen. If the time was long enough the crime could still have been committed, and the biological evidence could be gone. But what was the magic number of hours? In addition, for the male, he may have a disease or be vasectomized. He may have withdrawn or failed to ejaculate. There was no control for this. For the female, post assault activity, such as muscle movement in walking or running; vaginal bacteria; and deliberate activity, such as a shower or other cleanup, could be factors as well.

Someone had to ask those questions. If I did not write down the question, it probably did not get asked. For instance, picture a male police officer interviewing a young female on the details of a sex act. He was probably not going to get much due to the extreme embarrassment to both parties. If the questions were on the form, they would get asked. As experts on the witness stand, we were expected to evaluate the relationships between the witness statements and the evidence. We had to have the facts first, no matter how uncomfortable the question.

It obviously was difficult to assign a time between incidents and the survival of semen within the vagina. But some research had been done, and a number of papers were published on the subject. Within the literature, there was a direct conflict between sexual assault examiners and fertility study examiners. The results of sexual assault exams, which coincidentally matched my observations, showed about twenty-four hours before sperm were undetectable or the acid phosphatase test (AP) was negative on a vaginal swab. The AP test looks for an enzyme in high concentration in semen. In fertility studies, an extensive search for sperm might produce one or two sperm reportedly up to seventy-two hours later.

We would have discussions about this in the lab or later over a beer. Setting all science aside, was "love" the factor for the difference? One might think inactivity such as sleep for a woman prior to a fertility test might be a factor, but for seventy-two hours? That did not make much sense. We again rationalized that it was probably the quality of the examination and that perhaps fertility examiners just did a better job than we did. We had no idea why there was a difference in time. Eventually with the advent of DNA, the time of seventy-two hours was set as a standard for the eventual loss of sperm from the vagina of a woman after intercourse. Was it better skills or improvement in technology and instruments? We will never know.

I am sure any forensic serologist will tell you that the kind of conversations that take place within a crime lab over various sex acts and evidence expectations are not likely to be found anywhere else in society.

Someday someone should write a book of these conversations titled *Sexual Creativity and Evidence Expectations*.

Although I suspect that most folks probably would never get to the chapter on evidence expectations. If you can imagine it, it is probably written in a police report somewhere.

Early in my career testifying in a sexual assault case, I had one of those gotcha moments that all witnesses dread. I had found a pubic hair foreign to the victim collected in the sexual assault kit. Hair comparisons had value for exclusion but limited value for inclusion. The defense attorney, Ken Morrow, was probably the top defense attorney in Oregon. I knew him personally, as we both liked to fish. Occasionally I would see him on a riverbank fishing for steelhead and we would talk about fishing. In the courtroom he started questioning me about the frequency of finding pubic hairs in a sexual assault, and I went into an explanation of the various phases of hair growth and shedding and said that it was not uncommon.

"But how do you know?" he asked.

I said, "Well, it is obvious. If you look at the sheets in your own bed, you can find them."

Ken looked at me with a quiet smile while making eye contact, long eye contact.

I had just opened the door to my private life. Damn! I was had! He knew it, and I knew it. It was definitely an oh-shit moment. How did the pubes get in my bed? Did I have pubes? Were they there from previous sexual activity or not? Had there been anyone else in my bed besides me? I don't normally fidget, but I was fidgeting.

What about them? There was no end of questions he could ask from this point on, and I knew it.

My heart was pounding, and I could feel the sweat on my forehead while my mind was trying to find a way out of this predicament. Ken was silent, kind of savoring the moment and shuffling some papers in front of him, just watching me twitch in my seat on the witness stand and the perspiration accumulate. No doubt, I had the feeling something terrible was going to happen, and there was absolutely nothing I could do about it.

Then he said, "No further questions."

A few weeks later I saw him again on the riverbank. He looked at me and said with a smile, "I had you."

I said, "I know."

Nothing further was ever said about it. Some lessons in life, like this one, you cannot prepare for ahead of time but never forget.

In working up evidence, one occasionally runs into special circumstances for which there are no immediate answers. For example, in a sexual assault case where semen was found in a child's panties, a defense expert concluded that if the accused man was sleeping in the same bed with this child and she accidentally rubbed up against his private parts while he was having a wet dream, some semen may have been deposited on the exterior of her panties. All just by accident, and no crime was committed.

BS was the first word to appear in my mind, but it was a significant problem and no immediate solution. I could not think of any current procedure that addressed this problem. No doubt the panties were positive for semen, but was the deposition on the outside or inside? A new question, and what to do? I had to think about this for a while, and many cups of coffee later, I had an idea. When looking at a bloodstain, it was not uncommon to be able to tell which side of a cloth the stain was on. I just needed a way to visualize the invisible semen on these panties. It was no small task, even for the creative mind.

In the lab, we used a reagent called *acid phosphatase* for identifying the fluid fraction of semen. If semen is present, the test turns blue. I prepared some cloth samples from a semen control and made a gelatin mixture on a glass plate with the AP reagent within it. The gelatin substance was normally used for electrophoresis and prepared on an eight-by-ten-inch glass plate to a thickness of about one to two millimeters. I made two plates, placed the dry test cloth on the gel for a few minutes, and then turned it over and placed it on the other gel plate for a few minutes. The incubation times were equal for both sides. I removed the cloth and placed both gels in an incubator at thirty-seven degrees Celsius for fifteen minutes to allow a chemical reaction to take place, and then I removed the plates and added the coloring reagent.

This procedure made the invisible semen location visible. The diameter or the reactive area and intensity of the blue color would indicate which side was the original. I could outline the control stains with a marker using UV light to locate the stain and then compare the AP reaction with the original stain's diameter. One might ask, why not just use the UV light? There is a problem with UV. Urine contains a UV-positive substance called *Urobilinogen*. It was possible that the UV technique would give a false indication of semen stain size if any urine was present. The test worked. I could tell which side of the cloth the majority of the control stain was on. It took more time to fine-tune the procedure, and the evidence panties were tested in the same manner. The stain was on the inside of the panties; there was no doubt.

I know that today most scientists in crime labs do not have the flexibility to do an impromptu procedure like this.

But creativity and ingenuity are the foundations of science. Without them, are we truly scientists or just technicians?

In the 1980's, long before DNA testing, many times we had good semen stains or semen-vaginal mixtures but no suspect for comparison. With no data banks to rely upon, I would take a cutting of cloth or a swab from each one of these cases and place the evidence in a properly labeled envelope. The envelopes went into a box in the freezer.

The hope was someday there would be a way to locate the assailant. Over several years I was filling several boxes of samples from unresolved sexual assault cases and some homicides. Years later after I was promoted to lieutenant and managing another lab, I got a phone call asking if I wanted all of these samples; there were lots of them.

Apparently in preparation for an inspection, it was determined that this collection of evidence was not an approved storage procedure. Basically no one had written a procedure to retain samples from old cases quite like this. If I did not want them, they would be thrown out. I was speechless; we did not have the technology to work them up in the early 1990s, but someday someone might. What about the people who were victims of these crimes? I pleaded that they hide them until after the inspection and continue to hang on to them; maybe someday they could be examined and the perpetrators brought to justice.

That was the end of the conversation, and I did not hear back for a few years and assumed that was done. I was wrong. They were all thrown out. Looking back, if we had them today, the samples could have been run and compared to state and national data bases. Perhaps closure could be found for some of the victims by identifying the assailants through the data bases, but not now. Did the needs of the agency outweigh the needs of the community?

Was the decision to toss them out the right one? We will never know now. The contents of those boxes would have been a gold mine for today's cold-case investigations.

With the development of DNA testing in the late 1980s, I was sent to Cellmark Corporation in Rockville, Maryland, to learn the DNA-testing procedure. At that time, it was not very sensitive and required a bunch of sample to get a result. The samples were placed in a gel and an electrophoretic run made the procedure kind of fun. In the final steps, a radioactive reagent was added; it would remain at the site of each DNA marker. But the concentration of the DNA was so small that in order to see the results, the plate required being overlaid with an x-ray panel and allowed to incubate with the radioactive material to create an image on the x-ray panel. It might take weeks to see a result.

We all knew this was going to be the future in forensic serology and were looking forward to it. But after I had run several of these, the vision of doing this all day long and nothing else was not appealing. One day during the lecture phase of the class, there was a discussion on partial patterns that disturbed me. In essence, a total pattern in a DNA test could have eight or more markers, but what if some of the markers did not visualize? Each marker position on a plate had been studied in various populations around the world, and a frequency of each band could be assigned. The multiplication law of probability states that if events occur together but are independent of each other, the frequency of each of their occurrence can be multiplied. The final number is the frequency that the two independent events would occur simultaneously in nature. With multiple markers, such as we see in DNA testing, the numbers get astronomical, and frequencies of the random occurrence of all independent markers gets down to one person in millions of people.

But what about partial patterns? What if you only detect five or three? Are the implications still relevant? The company thought that they still were relevant and did the same calculations with whatever number of markers showed up. I was troubled with this concept, because what if one of the undetected markers was different? It only takes one nonmatch to be an exclusion! Is it ethical to give the death penalty to a person based on a partial pattern? In the multiplication law, because some markers match the accused, that does not give a higher probability that the next one will match, too. What if the accused had a brother who actually committed the crime? If the mom and dad are the same parents for all, some of the markers between siblings are going to match.

At no time did anyone advocate that a certain frequency would constitute an identification, but it was certainly implied. To me, not addressing this was an unnecessary bias. I argued that if you are going to use a partial pattern in a case, you must also state in the discussion of the report that one of the undetected markers could lead to an exclusion if detection of that marker had been possible. I was shot down but not without saying my piece. When I got back to Oregon, the word got out that I was argumentative while at their facility. Yup, I was.

My division director understood where I was coming from, and we moved on. He decided that DNA testing would only be done in one lab and wanted me to transfer to that lab to lead the team, despite my questionable performance at Cellmark. I declined. The work was somewhat interesting but extremely repetitive. I chose to stay where I was and seek promotion to a lab director position when one became available. The director could have directed me to transfer, but he never did. In our talks over coffee, I think he was a little like me; he missed fieldwork himself in his current position. I never regretted the decision. In 1990, I was promoted to lieutenant and director of the Coos Bay Crime Lab.

Chapter 10 Offering a New Science to Game Enforcement

The OSP was divided into several divisions, such as traffic, game, and forensic services, among others. All of us were sworn, and theoretically I could have transferred out to one of the other divisions and maintained the same pay grade. The patrol office was right next door to the crime lab, so we were in constant contact with the guys and gals on that side. My hobbies included hunting and fishing, so I was in contact with the game officers occasionally to see where the fish were biting. The lab did some casework for them in terms of determining if a meat or blood sample was Cervidae (deer or elk) in origin, and they also had the capacity to test for a number of other animals as well. They would come in, get a cup of coffee, and talk about cases and problems in getting convictions.

Each lab did some basic biological testing as well as matching bullets or cartridge casings back to a weapon. It was not uncommon for someone to poach a deer or elk, take a hind quarter, and leave the rest. At times a carcass would not be located for weeks and was quite odiferous by the time it was found.

Despite the smell, the officer would use a metal detector and try to locate the bullet somewhere inside. It was dirty work for such a small prize.

I was known as the go-to guy if you had an unusual problem. One of the first problems I addressed for them was seeking a method to separate deer meat from elk meat. It was not uncommon for an officer to serve a search warrant on a freezer and remove several packages of hand-wrapped meat. At that time, the fine for poaching a deer was about $250 and about $1,500 for an elk. The courts always applied the lower penalty if there was no certainty in the origin.

I started the hunt by running a procedure called *immunoelectrophoresis*. As I stated before, this is a process in which proteins are exposed to an electrical charge and a separation of the various proteins takes place on the gel support medium. I confess, as before, that is a very simple explanation for a complex process.

After the run was complete, an antiserum to the Cervidae family(deer) was applied to the plate, and the proteins form visible bands where there were antigen-antibody reactions. The location of the various proteins on the plate was specific to a species, but there were bands that varied from animal to animal within a species as well; these were a blood type for that species. Given enough time, I could probably identify each protein, but I needed something simpler.

While at a forensic conference, I met a scientist from the Alaska crime lab, and he was trying to separate moose and caribou meat. We compared notes, discussed possible solutions, and went back to our labs to do more work. I tried several polymorphic enzymes common in all mammals and came up with one that had promise. It was called *PGI*, or Phosphoglucose Isomerase. I ran several deer and elk samples, and based on the protein migrations, I could separate the two species as shown below. Now I needed lots of samples.

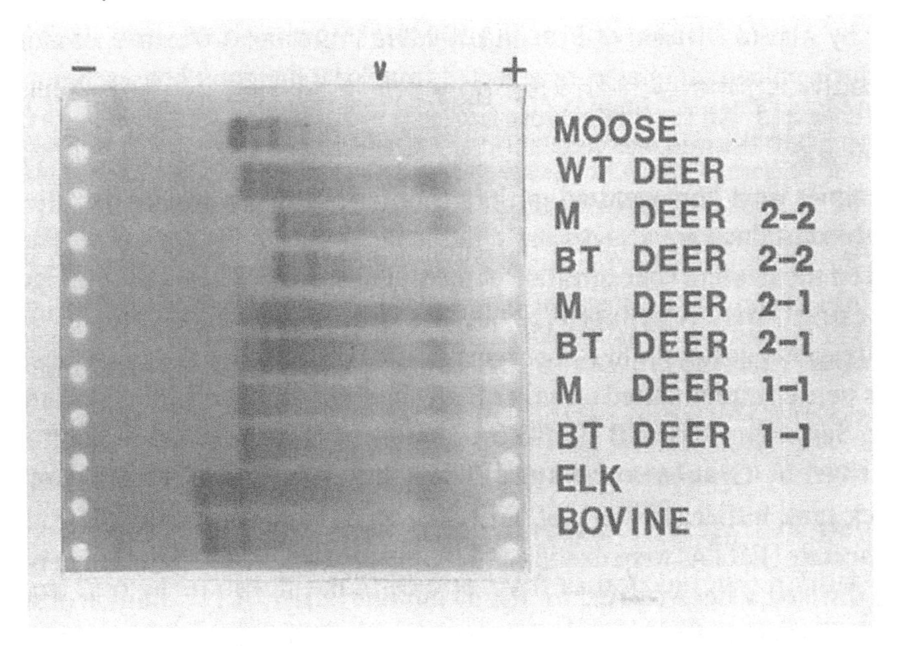

Figure 9

I contacted the game sergeant and told him I needed lots of samples from all over Oregon. I asked them to just take a page out of their notebook, smear it with blood when they were in contact with a dead animal, identify the location, and send it to me by mail. In the first week of deer season, I had about two hundred deer samples and later got about that many elk samples. Here I had a big surprise: Not only could I separate deer from elk, but also there was some variability in the deer samples that amounted to a blood type for deer. The differences were genetic. My friend in Alaska was having similar success with his species testing, and we jointly published our findings in the *Journal of Forensic Sciences*.

My boss was so happy that he made me the liaison between the labs and the game division, and as such, I had to go to their in-service schools. OSP was heavy on the cheap side, and they always held these meetings in old military facilities. These were usually Quonset huts left over from World War II that slept at least thirty or more people in one open building. It was great for conversation, but I am a light sleeper. As soon as the lights went out, the snoring started from at least twenty-nine of the thirty guys there. I sat on my bed in misery; I swear I could hear the sheet-metal roof rattling. I would gather my bedding, take it out to my car, and sleep in the back seat. If you have ever tried this, you will know that there is no comfortable way to sleep, but it was quiet.

If the meeting lasted a week, there was another problem. All the guys showed up in marked pickups. I had the only unmarked vehicle on base. Wow, did I make new friends. One night we packed the sedan with all the guys we could get in it and headed into downtown Portland for a little evening excitement. We hit some bars and ended up in northeast Portland, traveling down a street headed for another dive. For some unknown reason, the headlights went out. I pulled over and found that a fusible link related to the headlights on the fire wall inside the engine compartment had overheated, so we waited for it to cool off.

This was taking some time, and we were getting worried. Vehicles with dark-tinted windows kept driving slowly by us, and no one had a firearm. That was probably the upside. Finally, a couple of these vehicles stopped next to us and just sat there, and we could not see the occupants. I wasn't sure where we were, but I said it was time to leave! I started the car, still without headlights, and took off down the street at much more than the speed limit, with my foot pushing the gas pedal to the floor. I am not sure how fast we were going; I was not looking at the speedometer. People were flashing lights and honking at us, but there was nothing I could do; I was not slowing down. If only they knew it was a car full of cops breaking most of the traffic rules in town! Just think how that would look in the *Oregonian*. About a mile down the road, the lights came back on, and we hustled back to base before we were apprehended. Everybody bailed, and I climbed into the back seat for the rest of the night.

My attempts at helping the game division were not always a success. They had cases in which people were snagging salmon and would cut the line just before the game officer got there. So I did a study on fishing lines looking at diameter, color, and extrusion marks under the microscope. This seemed like it was a good idea, but a company might make several miles of fishing line before it changed the dies the nylon was extruded through. I just could not find uniqueness in fishing line.

Back on the poaching issue, game officers might stop a vehicle at night, and the vehicle would have a deer carcass in the back. It was illegal to shoot deer at night, but deer had a flaw in their actions. If you shine a bright light on them, they will freeze, making them an easy target for a poacher. No one had ever developed a reliable method for determining how long a deer had been dead. There had been some studies on body temperature, but these were not reliable for court. Could anything be done?

In my review of the literature back in the 1980s, I found that trying to determine time of death in people was a hot topic. Eye fluid, called *vitreous humor*, was usually the fluid of choice in attempting to measure changes after death. This is a sealed fluid that is easily accessible. Getting Blood for proteins and electrolytes were never found reliable.

I read all the literature I could find and then contacted some of the published authors to get more detail on how they were doing their experiments on eye fluid. I soon found that some of the work done at that time by health professionals was unreliable. As an example, a forensic pathologist at a teaching hospital would assign a med student or intern to collect a blood sample or eye fluid sample from a recently deceased person. The time of death was available in the charts, but the time of collection was not recorded. These guys would get to it when they had the time. Some biological markers literally change by the minute in the first twenty-four hours after death, but somehow this was overlooked. Also noteworthy, these samples were all collected from sick people. The very enzymes and electrolytes they were searching may have been abnormal before the person died. If I was going to make this work for game cases, I would have to go with eye fluid, and there had to be certainty in the time of death and collection.

The hunting laws required that the head of an animal had to accompany the rest of the carcass during transport by a hunter after he or she bagged an animal, a fortunate circumstance. There are two fluids in the eye: the aqueous humor in front of the lens and the vitreous humor behind the lens.

The aqueous humor was more like water and easier to get into a syringe. My coworker Kenn was working with me, and we settled on this fluid for our testing. I met with the local medical examiner and explained what I was researching. As usual, he was very supportive and advised the hospital lab to assist me. There I tried testing a number of compounds, such as lactose, and elements, such as sodium and potassium. I finally settled on glucose. In addition, there had been a limited study done in Washington state that suggested that deep muscle temperature could also be useful. There was a concern that the weight of the animal would affect the decline in temperature as well as the outside temperature. Only working with the animals would resolve that question.

We got some specimens from the game officers with documentation regarding time of death (TOD) and time of collection. In these few initial samples, we saw a linear regression in glucose values with time up to about eight hours. It was apparent that after the brain and heart stop functioning, some tissues live on until their fuel source or waste products eventually kill them. Glucose was a fuel source and declined as the surrounding cells consumed it. Kenn and I reported our findings to division, and they gave us all the time we wanted, as long as it was off duty. Our caseload had to be maintained. We were not to be deterred. We needed specimens and went on a campaign to get samples.

We heard about a special hunt in the upper Willamette Valley. The entire twenty square mile hunting area was completely fenced, and there was only one road in and out. Oregon State University students were already doing a project sampling all the deer that were shot regardless of gender. The weight and health of all the deer examined were recorded. The hunts occurred on weekends in November, which was just right for us, despite meaning that we missed the college football games.

I made up a special tag that the hunter would use to list the time of the kill, and we stood at the gate handing them out as the hunters went in. Keep in mind, we had to drive an hour and be there before daylight each day. The hunt went on for four weekends each year, and Kenn and I were there all day, each day, to receive the tags, record and collect a deep muscle temperature, and get an eye fluid sample. Sometimes the hunters did not come out until after dark which made for some long days. We did this every November for three years and ended up with eye fluid samples and deep muscle temperatures from 240 deer. We ran the glucose test in the lab ourselves and correlated the results to the deep muscle temperatures that we obtained on site. The results were encouraging; it was working. Any time we had a question about a specific animal, we could call the university, and they had all the details on that animal available.

We had a number of concerns in the study, such as bullets to the head, animal size, outside temperature, and methods of dressing the animal just to mention a few. We sent all our data to the U of O math department for analysis and independent review. We got back a linear-regression model that was highly reproducible from animal to animal. Based on our concerns, we set limits on deer weight and limited it to the Willamette valley and blacktailed deer. Our confidence interval for the first eight hours was plus or minus about one hour. That meant that if we got a time of 11:00 p.m., the reliable range was between 10:00 p.m. and midnight. Eight hours after death, the glucose value was usually zero. The hunters had to be given the benefit of doubt if the TODs occurred around dawn or dusk. That was ok.

We had done the research backward by doing the field study first. After we saw our results, Kenn and I knew we were on to something no one else had ever done. That was, getting a reliable estimate on how long any species had been dead. We were still working for free, but we were excited about our results so far.

All that was left was a control study. This is where you shoot an animal and stay with it to collect samples every hour for the entire first eight hours. We needed at least twenty animals. By doing it ourselves, the results should be even better.

We got word of a wildlife veterinarian in Corvallis who held a special permit to shoot blacktailed deer for a disease study. He was a bit eccentric, but Kenn and I liked him. We jumped in an old van with him and were off to do some legal spotlighting on a Friday night. There was no concern about falling asleep; his driving on old mountain roads and constant jabber kept us wideawake. He would tell stories about his big game exploits all night long. He would boast that he could completely gut a deer in thirty seconds; I doubted that until I saw him do it.

We tagged along with him. Each night he would spotlight and shoot several deer. We would get the initial data and transport them to a warehouse, where Kenn and I would babysit the deer the rest of the night until we had all the data we needed. This was no easy task. It took several nights to get twenty deer, but we got our data. We were sitting in a cold warehouse with dead animals just waiting for another hour to pass. We had been up all night and now most of the day, and we were beat by the time we finally went home. Did I mention we were not getting paid for this?

The data was turned over to the U of O again for analysis and proved again to be quite reliable with tighter confidence intervals. This study was uniquely adapted to animals instead of humans, as diabetes was unheard of in deer. Internal temperature variation was sufficiently rare also to not be a factor.

Kenn and I published our findings in the *Journal of Forensic Sciences*. It was the first time anywhere in the world that a reliable time-of-death method was published for any species. High fives were in order all around, especially in the game division. It was especially nice when we received publication requests from other countries and calls from other researchers looking for more information. I'm not sure our families agreed, but we thought the time was worthwhile. We were now ready to turn this over to the game division.

Kenn and I built a TOD (time of death) kit for the game officers and started handing them out. They contained instructions, two tubes for eye fluid, and a syringe and needle. We preferred if they took two samples an hour apart so that we could demonstrate that the change in glucose and temperature values were consistent with the study. To our surprise, we did not see many kits coming back. To everyone's surprise, when the officers confronted suspected poachers, they would say they shot the animal during normal hunting hours, a not unexpected response, and that's normally where these cases ended. But when the kits came out and the officers explained what the lab could do for them, the poachers usually confessed. These people were usually not hardened criminals familiar with the workings of the law. So when faced with the potential truth, they gave up. This was happening with regularity and explained the absence of kits coming back to us.

The number of cases made based on the potential for testing was exceeding the number of cases where testing was actually performed. When word spread around the state, the number of nighttime hunter contacts went down.

Kenn and I were a hit at headquarters, and there was talk about all kinds of applications. Our division boss thought this was so good that he wanted to expand it to mule deer, both species of elk, and also moose. Theoretically there was no reason that the test would not work, but not without a control study. He thought that was a waste of time. *Just go with it on all these animals*, was the directive.

Kenn and I knew this could blow up in court and the entire study could be placed in jeopardy, so we said no. I'm not sure where we stood at that point, but it was a risk we had to take. We were careful in limiting our results to within specific weight limits and temperature environment of the study just for blacktailed deer. If we expanded the use beyond the research, we risked losing the reliability of the entire study. HQ eventually acquiesced but never allowed us to expand the study. The word around there was that we just wanted to get some hunting in on government time. For all the time Kenn and I volunteered away from our families, this was how much trust they had in us?

In later years when Kenn and I left the lab system, the method was discontinued. No reason was given.

The new folks running the labs were nonsworn personnel and did not grasp the importance of this test to the game enforcement community. A lot of work was lost to history. As I reflect on the study, it would be so easy to reintroduce the method by using handheld glucose meters with test strips. Once proper validation was done, the game troopers could do the test themselves. Opportunity for change is still present.

Moving on and many years later, my son and I were bow hunting elk in late September in the mountains of western Oregon. There was a certain solitude that normally comes with bow hunting. There was the early morning cool air, the fresh breeze, and the sound of a bull elk bugling in the distance. We dressed in camo and face paint, had bushed ourselves with cedar boughs to hide our scent, and at that moment were approaching an elk herd deep in the forest. The wind was right, and it could get exciting real fast. Then it happened...Bang! A rifle went off not a quarter mile from our location. The elk heard it, too, and were gone in an instant. The Oregon Department of Fish and Game, in their wisdom, had scheduled a rifle bear hunt the same time as bow season. But I suspected that this was not a bear hunt. I had heard far too much shooting in the last week to know it was all for bear. They were just too hard to find.

Later that day I talked with one of our local game officers, and he confirmed my suspicions. People were illegally hunting elk with a rifle and then sticking an arrow in the bullet hole to make it look like an arrow shot. They did not have a good idea about how to get around this practice. I was ticked off and was going to do something about it, just not sure how I was going to approach it yet. But I would find a way.

I knew enough about firearms and wound ballistics to know there had to be a way to fix this. When a person or animal is shot with a bullet, the bullet normally breaks up to some degree depending on its design. This leaves a track of bullet fragments all along the wound canal. In humans, we call this the *snowstorm effect*. The photo shown in Figure 10 below demonstrates the lead particulates along the path of a lead bullet in ballistic gelatin. I could spot this in a heartbeat, but how do you get an x-ray when you have a deer or elk in the field? These particulates are normally too small to carve out with a pocket knife and the suggestion of taking an animal carcass to a medical clinic for an x-ray got plenty of laughs from the game officers.

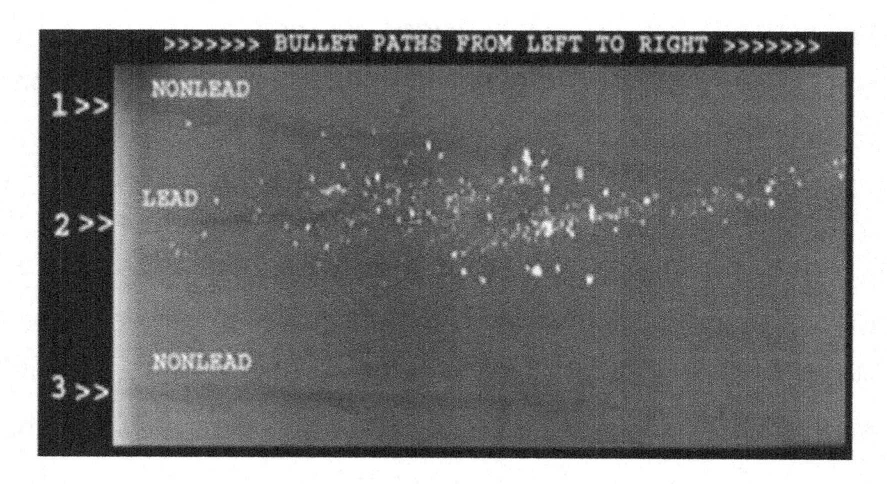

Figure 10

That left veterinarians, but they wanted money for an x-ray and consultation. I knew the department was not going to pay for this service. If the meat spoiled while the officer had it in his possession, the department was liable too.

I knew that many of the local police agencies had received portable x-ray machines through a federal grant. The machines were set up for bomb detection but were rarely used. Typically, they might get used once every couple months in the larger agencies. They were portable, battery operated, and used a small Polaroid film pack for visualizing the item screened. The Polaroid film was only about three inches by four inches in size, too small for what we needed. I went back to a local medical clinic and asked if they would at least provide us with unexposed x-ray cassettes and later develop them for us. They said they were happy to cooperate and provided the materials.

I set up the device in the lab, learned how to run it, and adjusted the distances from the lens to the target to get the best image. It took three or four attempts, but I got it about right. A few days later, one of the game officers called in and had a road-killed deer in the back of his truck. He brought it up to the back parking lot of the lab and I shot it in the head with a handgun. We x-rayed the head and got a great image of the snowstorm effect. Hunting arrow tips are made from steel and do not fragment on entry. We had a tool now to make a difference. Just before the next season opened, we did an article in the newspaper about what we had developed, and it was the talk of the town. Figure 11 shows the portable x-ray.

Figure 11

That next season my son and I were again hunting elk in the Cascade Mountains of western Oregon. We did not get an elk. But it was beautiful country, and it was good to be there. Not a shot was heard all day long. It was amazing how all those bears had disappeared. Not to mention that I had vindication for the previous year.

In my last contribution to the game division, I took on the issue of cheating on a salmon tag. In a modern world where hunting and fishing is highly regulated, the salmon are probably the most controlled of any species.

The specific species, the number per day and year that can be taken, status as clipped or unclipped, and minimum size are all factors that dominate the sport. The ODFW have licenses and tags available that you have to buy in order to catch fish. When you are a fisherman, once you navigate the legal side of when and where to catch a fish, there is the sport itself. When you are so lucky as to catch a salmon, you have to complete the information associated with the catch on your tag. Not doing so results in a fine. After so many fish, you can't fish any more that week. If you fill the tag, you are done fishing. So if you are getting really lucky, what do you do but alter the date of the catch on your tag. This was working well for some, and the game officers were helpless in doing anything about it.

One day a game officer brought in a tag he had confiscated and was certain there was at least two different inks used on the date. I laughed. I could handle this one; I remembered my early days as a forensic scientist when I used to do ink comparisons. Now we had new tools, a machine called an *alternate light source*. I could shine different wavelengths of light on the ink and see different colors reflect depending on the chemical composition of the ink. Picking out two different inks and generating a photograph showing the difference was child's play if they used two different pens on one date. The technique worked flawlessly. Again, I was a hit with the local game officers, and they were catching one of these guys about every week during the salmon season.

There was just no limit to the amount of fun you can have at the expense of the people who violate the law. Sometimes I had to pinch myself to believe I was getting paid for this. This was just too much fun.

Chapter 11 A New Method in Testing Mushrooms

Every crime lab is tasked with drug testing, and ours was no different. There were times when drug testing took up the entire day or days. It was OK. It was an opportunity to refresh my chemistry skills and occasionally come upon a compound that eluded identification, at least initially. It was challenging to examine the instrumental data and try to determine what you had before you.

Natural substances, such as mushrooms, offered a change of pace in the day. First, you determined if they met the visual test. Then came the extraction process. Like all plants, there were lots of compounds present, and it was my job to find the one that was controlled. Psilocybin mushrooms were normally dried by the user, placed in hot water like a tea, and consumed. The reported effect was a psychedelic high similar to LSD. There were two controlled substances in these 'shrooms. One was psilocybin, and the other was psilocin. The psilocybin was in much greater concentration, and both compounds had complex structures but differed only by a phosphate radical hanging on the psilocybin molecule.

Testing required first identifying the psilocybin and then, through a lengthy process, knocking off the phosphate radical, thus manufacturing the second compound,

psilocin, as part of the confirmation. The test was time-consuming and at times only minimally successful.

One day I was sitting across from one of my associates who had an MS in organic chemistry and was considered the authority on all things chemistry related in the lab. He certainly was into this stuff far more than I was. But my degree also included organic chemistry, and I could usually hold my own in a discussion. One day we were lamenting the problem of getting that phosphate radical off the psilocybin molecule and discussing the literature reports, knowing we were going to have to perform on of these procedures before the day was out.

My associate was a thinker. He had thick glasses and always looked the mad scientist with a pencil protector in his shirt pocket. He was never without his briefcase and his calculator. I secretly believed he would spend his evenings looking up words to use on us, while I spent my evenings thinking of tricks to play on him. Sometimes when he was expounding on a topic and then walked away, we would ask each other, "What the hell did he say?"

In the human body, there are a number of ways the metabolism of life deals with phosphate radicals. One common pathway includes the enzyme acid phosphatase.

All cells have a little, but semen contains a lot, and as mentioned earlier, it is used as a test for seminal fluid also.

I mentioned that it would be interesting to see if we could cleave that radical with seminal fluid incubated with an extract of the mushrooms. My guy thought about this some, got up, and walked to the restroom with a cup. Are you kidding me? No matter how bad I had to pee, I was not going down there. Was he doing what I thought he was doing? Sure enough, he returned a short time later with a specimen that we shall call *reagent A*. Well, I brought it up, so here we go.

I separated the cells from the seminal plasma and prepared a dilution that should work. I mixed some ground-up mushrooms with the reagent A and allowed the concoction to incubate for a while. In the meantime, I put a drop of the cellular fraction on a slide and placed it under the microscope. My associate looked first and observed cells active and swimming as single-celled flagellates.

I took a look and shouted so all could hear, "My God, there is one in there with glasses and a briefcase!" He was caught off guard and rushed to the microscope to look before he realized I had got him...again. What joy, I could hardly control myself.

I took a sample of the mushroom concoction and ran it through the testing process. The results were amazing. I had removed nearly 100 percent of the phosphate radical in my first attempt. There were high fives all around, except for our chemist; no one wanted to touch his hand. I ran some more tests and determined the appropriate concentration and incubation time for optimal results. We quickly wrote up a memorandum and sent it out to the other labs as a new procedure. We were quite proud of our work, a rather "handy" breakthrough in forensic science.

It did not take long for feedback. The brass at headquarters were not scientists, nor were they considered liberal in their thinking. My boss got a phone call from headquarters, something about police officers involved in questionable activities. What would the public think if word got out? With that, our new science was discontinued, and it was back to the old methods. The concept of someone in each lab collecting reagent A, while on duty, did not sit well with HQ. The other labs and their chemists were impressed by the technique, but no one was running down the hall for reagent collection. We would occasionally get these short messages addressed to my associate asking if he enjoyed his work a little more these days. Or they would just say they wanted to give him a hand for his accomplishments. That would usually wind him up for the rest of the day.

It was a scientific breakthrough in forensic science, but it looked like it was going to have to wait for another time and a more open-minded administration. That didn't happen.

Chapter 12 The Diane Downs Case

In May 1983, I got a call for assistance on a homicide in Springfield, Oregon. I had been on the force now for five years and was capable of handling routine call-outs by myself, if needed. This one turned out to be a little different. Three children had been shot as well as the mother, late in the evening on a rural roadway near Mohawk, Oregon. The mother's name was Diane Downs.

Good Morning

Eugene Register-Guard

Sunny

Search continues for murder suspect

Two youths remain in critical condition

Figure 12

Reportedly the three children had been shot in the vehicle, and the mother was shot standing in the driver's-side doorway.

The mother reported to police that she was returning from visiting a friend and traveling down a country road. There she was flagged over by a man with dark hair. She stopped the vehicle and asked what the man wanted. She reported that the man had a gun and wanted her vehicle, and she refused to give it to him. She said she faked throwing the car keys in her hand, and he apparently believed that the keys were gone. He shot her in the left forearm, pushed her out of the way, stood in the doorway driver's side, and shot all three children without entering the vehicle.

Ms. Downs stated that after shots were fired, she pushed him out of the way and rushed to the hospital with her children. At the hospital, all the children were rushed inside for treatment. Unfortunately, the child on the front passenger floor was dead. The other two children were alive but in critical condition. When there was time, Ms. Downs was also treated for her injury. Shortly after the arrival of the Downs family, a sheriff's deputy was called to the hospital to take a report from Ms. Downs.

Around midnight I met John from the SO at the county shops to process a red Nissan sedan. This was a newer compact vehicle with a light-colored interior and in good condition.

As soon as we turned on the bright lights, John and I located some .22 cartridge cases in the vehicle, which gave us our first clue that the weapon used was a semiautomatic in .22 caliber. For clarification purposes and clarity in terminology, unfired ammunition is called a *cartridge*. When a cartridge is fired, the bullet goes out the end of the barrel, and the remaining case is in the chamber of the weapon. Semiautomatic pistols eject a cartridge case after each discharge, and in a perfect environment, they will usually cast the casings to the right. There are lots of variables on where a cartridge case could end up, from bouncing around inside a vehicle was certainly one of them. Downs's shooting statement was consistent if a person was standing in the driver-side doorway and firing the weapon with the hand reaching inside. There were also bloodstains in the back seat area that included a bloodspatter pattern on the ceiling and rear window near where the oldest girl was sitting. We will come back to that later.

On the passenger front console, I located a small lead smear consistent with a bullet impact site. I could not find the bullet anywhere on the floor or under the seats, so I began removing the carpeting. I found a .22 lead bullet under the dash and under the carpet in front of the passenger seat. If the bullet had struck this area unimpeded, there should have been more damage to the console, a suggestion that the bullet was slowed by an intervening object, which could have been the child in the front seat. With the casings in the car and a .22 bullet, we now knew the murder weapon was a .22 semiauto handgun.

We finished up about daylight and attended an investigator's meeting to hear any updates, and then I went back to the vehicle. I always like to see outdoor evidence again in the daylight. Now having a better understanding of the statements made by the mother on how the shooting took place, I got a mechanic's creeper and went all along the outside of the vehicle with a bright light source just to look again. There is a common phrase in our business, *I am not sure what I am looking for but will know it when I see it.* It could not have been truer than at this moment. As I passed the front passenger's door, I saw some small red dots on the rocker panel that looked like blood. Figure 13 and 14 show strings that were based on trajectory calculations from representative drops of blood. The stand in Figure 14 shows the location where the strings converged.

I did a quick presumptive test for blood, and it was positive. Then I opened the passenger door and saw bloodspatter on the threshold. The door had to be open when this spatter was generated, and the spatter originated from outside the vehicle and can be seen in the photo, Figure 15, next to the arrow.

Figure 13

Figure 14

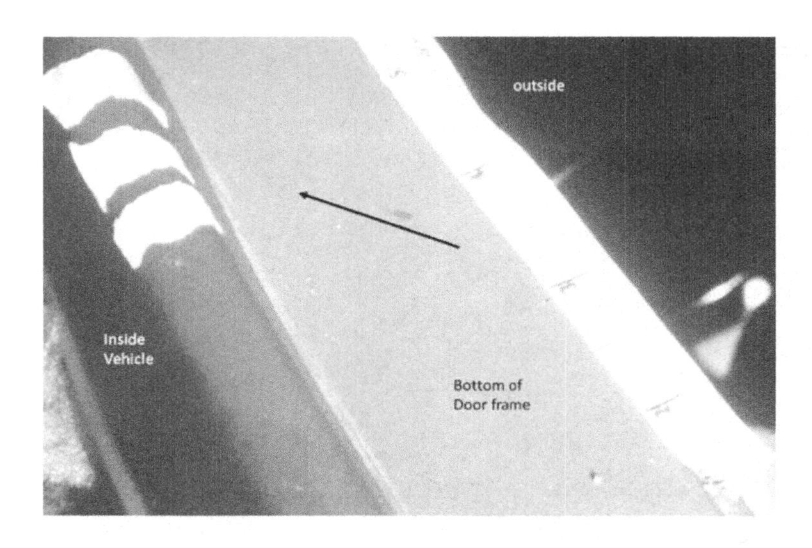

Figure 15

As is common with crime scenes, I was not sure what all this meant, but I was sure that the importance of this blood pattern would flush out with time. Was it human or animal? Did she run over something? What were the pattern expectations if she had run over something? I took a number of photos and arranged for the rocker panel to be carefully removed and preserved. I reported the finding to the investigators but cautioned them on the meaning until I had more time to work with it. This was exciting; I was pretty sure that someone was shot outside the passenger door, but I still remained cautious.

Taking the rocker panel back to the lab, I got a closer look under the microscope. The droplets were one to two millimeters in diameter, and some showed an upward trajectory. How could this happen?

Could you create a pattern like this from a moving vehicle? Were there possibilities out there that I had not considered?

I showed what I had to the other guys in the lab and then tested one of the small stains for human origin. It turned out to be human origin. All of us knew what a pattern from a moving vehicle would look like; all you had to do was drive through a mud puddle and look at the fender behind the wheel. The pattern of mud creates long-direction droplets that show a common origin from the wheel. These did not. The origin was directly in front of the pattern.

At this point it was decision time about what to do as far as blood typing. Leave the pattern intact to show a jury, or consume the pattern in testing hoping for a conclusive result? The tests available to us in the early 1980s used more blood than I had on hand. Also the victims in this case were all related, making it more difficult to separate them. The group determination was to leave the pattern intact. In criminal cases, the defense has the right to examine the evidence and have it tested independent of my testing. If I consumed all the evidence, the whole thing might be thrown out of court because the defense did not have a chance to test it, too. At this point in time, No suspect had been arrested, so there were no defense experts.

I measured the length and width of all the blood drops on the rocker panel with a micrometer scale and calculated their angle of trajectory when they struck the rocker panel. The calculations showed that the origin of the droplets was nine inches above the ground and about twelve inches away from the panel. Also to be considered was the size of the individual droplets. We call them *impact spatter*, and you can only get droplets this small in a shooting, or from blood dripping into liquid blood, or coughing blood. In our profession, for coughed blood, we used the fancy term *expectorated blood*. I had been to the scene, and a shooting pattern origin as described by Diane Downs was too high off the ground and was eliminated for all practical purposes. In order to have coughed blood, the person has to have blood in his or her mouth. The dead child did not. Blood dripping into blood was excluded also.

When I was done with the vehicle a second time, I drove to the hospital in time to attend the autopsy of the girl from the front seat. The morgue was a stuffy place, cold and without much ventilation, in the basement of the local hospital. No one normally goes down there in the dungeon, for obvious reasons. Dr. Wilson, the medical examiner, was well respected by attorneys and law enforcement throughout the county. He had been a mentor to me for the past several years, as we had much in common. He was tall, had dark hair, and was the constant student of his trade, always researching, always studying the literature for new science.

In my role with OSP, we were required to attend all autopsies associated with our cases, so we spent a lot of time together. Sometimes, at the beginning of an autopsy, he would say, "Let's see what we can learn today." It was always a great way to start. In this instance, the girl had been shot twice, and the angle of the bullet trajectories was interesting in that they had origins on opposite sides of the body. One must consider that there could have been movement of the child or the shooter in between shots. One bullet was in the car; the other bullet was not found at the scene. It was still in the child.

When the clothing was removed, I could see stippling around the bullet-entry holes on her skin. This meant that the barrel of the weapon was sufficiently close that the unburned or partially burned gunpowder had exited after the bullet had penetrated the skin around the bullet holes. When a weapon is discharged, especially if it has a short barrel, not all the gunpowder has a chance to completely burn; it passes out the barrel after the bullet. If the target is close enough, the unburned or partially burned particulates will penetrate the skin.

In addition, the size of the ring of unburned powder is proportional to distance from the muzzle. This can be reproduced, if the weapon is available, by shooting into paper targets.

I explained what I had seen on the vehicle and how things were not adding up based on the information I had from the mother. The shooter would have had to have awfully long arms to reach clear across the vehicle and have the bullet trajectories that we were finding. At this point we knew that the weapon was a semiautomatic .22, that someone was shot outside the vehicle, and that one child was shot from opposite sides of the body. The end of the muzzle was too close. I did note that she did not have blood in her mouth, an important factor related to the spatter on the rocker panel. Also, we knew that the passenger door was open while blood was spattered.

Autopsies, especially on a child, are sobering. I left the hospital and wanted to go home and give my children hugs. But just as I was leaving, I was requested to aid in processing Ms. Downs's apartment. Several of us entered the apartment with a search warrant and conducted a search. It was a duplex apartment, and the interior showed the common attributes of toys, clothing, and furniture that could be found anywhere in town. During the search, a .22 rifle was located under the bed. I happened to be standing by the bed when one of the investigators located it and placed it on the top of the bed. There it was photographed, marked, and packaged for transport back to the lab. Everyone wondered if this could be the murder weapon. In crimes in which children are involved, parents are often considered suspects until they can be eliminated.

Subsequent examination would reveal no cartridge in the chamber but nine cartridges in the magazine. The total capacity of the rifle tubular magazine was around fifteen cartridges. I am not sure when I knew it, but there had been talk that Ms. Downs had owned a .22 handgun identified as a Ruger pistol. I knew that the magazine of a Ruger pistol would hold nine rounds. I also knew that it is not uncommon for folks to move unfired ammunition from one weapon to another; I have done it myself. We collected the rifle and the ammo and were back in the lab by the afternoon.

I was tired. But there was a rush to compare the rifle with the cartridge casings from the vehicle, so I had to run to a local gun store to secure the same brand of ammo. When I got to the gun store, they were doing a brisk business in gun sales. Word was out that there was a nut out there shooting children, and folks were buying up weapons. I could not believe what I was seeing. That evening the case made front page of the *Register-Guard* newspaper, and the whole world knew there was a child killer out in the community somewhere.

The local police departments had immediately placed a high priority on this case, and there was no going home.

The cartridges were carefully removed from rifle magazine, marked, and prepared for examination with a comparison microscope. Factory ammo that I purchased was test fired to provide cartridge casing and bullet standards.

The casings and bullets recovered were compared to the testfired evidence. For clarification, A cartridge is a complete round ready to be fired in a weapon. The cartridge has a bullet and a case. The case is usually a soft brass tube that holds the powder.

Although the cartridges from the rifle had not been fired, it was not uncommon to find toolmarks made by previous actions, such as running a cartridge through a weapon's action and ejecting them, leaving an extractor mark on the rim. Firearms are manufactured with hard steel components, and cartridge cases are a soft brass alloy. Anytime the softer metals come in contact with steel, there is often a transfer of scratches onto the softer metal. The steel components were manufactured to certain tolerances, and the components were shaped and ground in such a manner that the scratches created on the softer metal are unique to a firearm and can be reproduced time after time in the test-firing of a firearm. A cartridge also enters the action from the steel magazine, which often leaves marks on the rim before it enters the chamber.

When you pull the trigger, a sequence of events takes place that creates a number of toolmarks that can be utilized for identification to a firearm.

First, the firing pin strikes the cartridge case and makes a characteristic shape and often unique etching when observed under a microscope.
Then the priming compound and powder detonate, causing the pressure to expand the softer metal against the chamber wall and back toward the face of the bolt, creating more toolmarks. Last, the bullet exits the barrel, and the fired cartridge case has to be removed so that a new cartridge can enter the chamber. An extracting tool within the firearm hooks the rim and removes the fired case from the chamber until it strikes a flat blade or surface called an *ejector*. At this point, the cartridge case goes flying out of the weapon.

Every movement of the cartridge or case had the potential of leaving toolmarks that could be seen under the microscope. In automatic weapons, all this happens in an instant. If the weapon is a revolver, the cases would have to be manually removed. Most revolvers hold from five to nine cartridges in the cylinder. Although it is not a certainty, if you find cartridge cases at the scene, the case was probably automatically ejected from a semiautomatic firearm. The comparison of toolmarks on test-fired ammo from the rifle with the casings from the vehicle was a not match. This weapon did not shoot the Downs children.

Ms. Downs had described the location of the shootings. Late that afternoon my boss, Chuck, and I traveled there to meet with other law enforcement officials. This was a paved rural roadway not located near any buildings and close to the Mohawk River.

Two cartridge cases were found, photographed, and collected. We were not sure how to label this location, as someone could have tossed these out the window, so it was always referred to as the casings location, not the shooting scene.

Teams of officers and Boy Scouts were assembled, and roadside searches were conducted from this location in both directions for miles. Several of us later walked the entire distance from this location almost to the hospital, about five miles distant. While I was working at the casings location, a guy came up to us and shouted, "What the hell? My wife is scared to leave the house; she does not want me to go to work for fear there is someone out here who wants to kill our children!" What could we say? Others came up to the scene with similar concerns. We could not help much; we were already suspicious that the mother was untruthful, but that does not mean you abandon all other possibilities. What if our suspicions were later found to be untrue?

By the time I left this scene, I was dragging and was hoping to go home; I was already into a twenty-hour day. I got back to the lab, and there was a discussion on how to locate additional blood. It might be possible to luminol the scene if it was dark enough. I think I got my point across that I was beat. So we waited until late the next night, and a group of us assembled at the casings scene. It was springtime.

The moon was out, and the sky was clear. It was not dark enough for me to see the luminol reaction on test samples I had brought with me. It had to be so dark you could not see your hand in front of you.

Luminol is a reagent that will luminesce when it makes contact with certain other chemicals. Blood is one of those chemicals, so this is a presumptive test at best but very sensitive. Theoretically, it can locate bloodstains at one to one hundred thousand dilution. Old stains worked better than fresher stains, but no one seems to know why.

We were out of luck in getting it dark enough, so I suggested we get a large tarp that I could put over me while we sprayed the luminol reagent. Several folks positioned themselves around a plastic tarp about twenty feet by twenty feet, and I crawled under it. We started at the edge of the suspected roadway and slowly moved down the road as I sprayed the ground. This was funny, and I could not help but laugh. We looked like a giant turtle going down the road with me on my hands and knees underneath, but it worked. I was getting positive hits that had to be examined by flashlight or blood-testing chemical reagents. As it turned out, dead insects are also luminol positive, and I found plenty of them but no blood. The absence of blood does not mean nothing happened there, just simply that nothing was found.

This is a very important statement when there is often an expectation that we can find anything, no matter how small.

I had never seen a community respond like what was happening around town in the following days. Everyone knew that a woman and her children had been shot and there was a bad guy on the loose in the community. Ms. Downs was happy to be interviewed and responded each time the detectives asked to speak to her. She was also actively taking interviews from the media. The case was becoming a media extravaganza, and the community was looking for answers from law enforcement. This put the pressure on all of us. The community was demanding results.

I attended daily investigator meetings, was on the phone off and on all day discussing different scenarios and was going over evidence for weeks. During this time, I also was trying to maintain my current caseload, which usually was a backlog of about five homicides. These agencies wanted results too. As the days passed, this case was consuming my entire days. Headquarters wanted constant updates, the DA's office wanted evidence examined immediately, and investigators had a million questions. The what-if scenarios were running rampant. Besides the phone calls, I got personal visits from the investigators as well. I think they were just looking for a place to hide out for a while before someone threw another assignment at them. Entertaining these guys took time.

Finally, I was able to sit down and look at the evidence from the children and the clothing Ms. Downs was wearing at the time of the incident. I also got a chance to take another look at the rifle and the ammunition that was within it.

The rifle made characteristic marks on the test-fired ammunition when a shot was fired. The tubular magazine of the rifle had a port where you could load and unload the ammunition without running it through the action. Knowing that was possible, I was surprised to see extractor marks on the cartridges from the rifle. The ammunition had not been worked through the receiver of the rifle, so why were they there?

The comparison microscope is interesting, in that you can place a scene cartridge case on one side and a test fired round from a weapon on the other side and see both objects at the same time when you look through the eyepiece. The items can be moved and adjusted to determine if toolmarks from the weapon are similar, dissimilar, or a nonmatch. I compared the marks on the intact cartridges from the rifle to the toolmarks on the cartridge casings from the scene and the vehicle. Oh my! To my surprise, the extractor marks were identical! I had to stop and think about this for a minute. This was so unexpected. In Figure 16, there is a thin line at the arrow that separates the two images. One image is a cartridge from the magazine of the rifle, the other is from the vehicle.

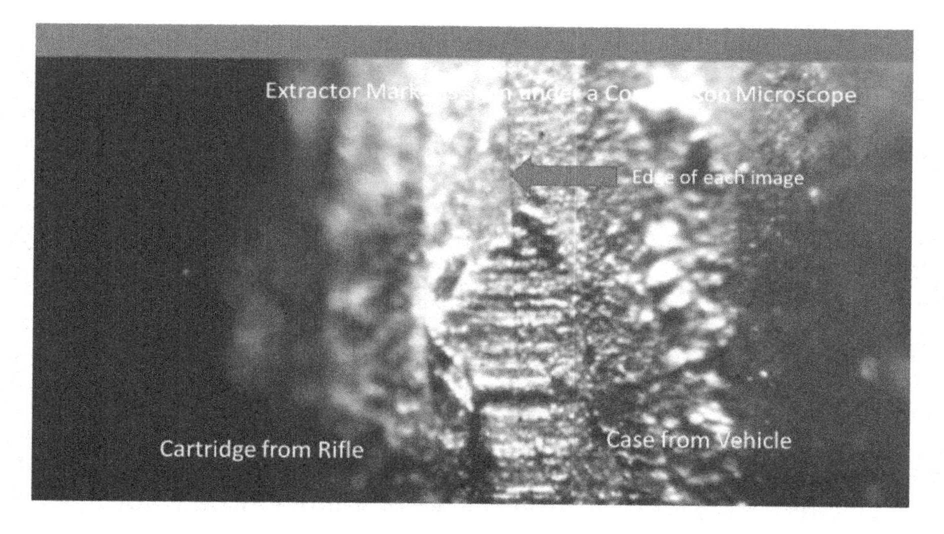

Figure 16

The casings from the vehicle and scene had been in the same firearm as the ammo from the rifle at some point in time. As mentioned earlier, it was known that Ms. Downs reportedly had another .22, a Ruger semiautomatic pistol. But that weapon had not been recovered.

We did some high-quality photography of the matching marks and passed the information immediately on to the DA and the investigators. Everyone was excited, but what to do next?

We now knew the weapon used was a semiautomatic .22 and that there was a relationship between ammunition in the rifle and the murder weapon. The plan was to keep this under wraps and continue to work the evidence. By this time, I was several weeks into this case and was getting underwater with my other cases. I went to my boss and told him I could not keep up with the daily demands of several other homicide cases as well as this one. He called Salem for direction, and it was decided that this case was a priority. I was removed from all other casework. This had never been done before in a crime lab that anyone was aware of, but headquarters was under pressure to keep this case a top priority. It was getting national attention, and everyone, including headquarters, was feeling the pressure. My associates would have to pick up the slack and add my cases to theirs. This was embarrassing for me, but I was running out of options.

Diane Downs did a video production of the shooting event that hit the national news. Her demeanor seemed unusual to most of us, and we knew there were facts in the video that were incorrect. But we did not give them up.

On the night of the shooting, she was shot in the left forearm and was treated and released from the ER. While she was in the ER, a deputy conducted a quick interview and obtained swabs from her hands to be tested for the presence of gunpowder residue (GSR). This is the test for trace elements of barium, antimony, and lead often found in the primer mix inside cartridge cases. These substances are microscopic and easily lost over time with use of the hands or handwashing. The test on Downs was negative for these substances, and she made a big deal out of this to demonstrate her innocence on national television. What she did not know was that the ammunition that was used in this crime did not contain barium and antimony, so it did not make any difference. Since she had a bullet wound of her own, the finding of lead was meaningless too. The finding of lead would be expected, as it forms a cloud on discharge and lands on anything close, including her hands. Since she was shot, one could not differentiate between her shooting a weapon and being shot by someone else with any confidence.

A few months into the case, I asked for a meeting with the ER staff. I needed more information than was in the police reports. These were the questions:

1. What was the actual position of the children in the vehicle?
2. What was the degree of consciousness?
3. What amount of blood in the vehicle was spilled on the pavement?
4. How did the children physically get into the hospital?
5. What did everyone see in terms of bullet holes and paths through the bodies?
6. Did they see any stippling?
7. What was the physical activity of Ms. Downs and the description of her injuries?

Last, I reinforced that we appreciated their cooperation. We all recognize that in an emergency situation, it is hard to remember some facts that are not important at the time, so I needed to see them as soon as possible before they would forget. What we learned was that Downs had gone into the ladies' restroom and washed her hands while she waited for her children to be treated. If ever there was GSR, this would have removed it.

Knowing that I would be doing proximity testing on the clothing, It would help to see the injuries to the living children and document them as well. When we got to the hospital, mom was standing guard outside the children's hospital rooms and would not let anyone in to talk to them.

Fortunately, the medical examiner has the right to conduct an investigation relating to the death of the other child, so I went with him to the rooms of the children and documented the injuries, much to the consternation of the mother outside the door. We may have had a search warrant; I just don't remember. Mom was raising a fuss while we were in the rooms, but there was nothing she could do.

The girl was sitting upright in her bed, and a nurse was in attendance. The girl was a sweetheart and reminded me of my own daughters. After a short introduction, the nurse removed the bandages so that we could see the wounds.

The bullet went all the way through her left hand, with the bullet entry on the outside surface of the hand, surrounded by a stippling pattern. I measured the size of the pattern. She also had two entry wounds into her left chest, one larger than the other.

Although there were three holes in total, there were probably only two shots. One shot went through her hand and was tumbling when it entered her left side. When a bullet is in flight, it turns like a properly thrown football. Once it strikes an object, it loses stability and begins to tumble. The next entry is usually oblong instead of circular. Bullets are usually heavier in the base than the pointed front. In order to keep the pointed end in front, the bullet has to rotate. If you stop the rotation, the heavier end wants to take the lead and the bullet tumbles. A hole created by a tumbling bullet will be larger than one that was properly rotating.

From the entry wounds, I could tell that she probably had her left hand crossed in front of her chest when the bullet entered the outside surface and exited the base of the thumb. The second shot was a clean circular entry. The Nissan is not a large sedan, but the distance required from muzzle to hand did not fit for a shooter in the driver's door and the child on the far side on the back passenger seat. The second shot to the chest was between nine and twelve inches. The bloodspatter was from the bullet through the hand and helped locate her position in the vehicle. If the muzzle is close enough to the hand, blood will spatter back toward the muzzle at the entry; this is called back spatter. It will also follow the bullet out of the hand on the opposite side; this is called forward spatter. Final results of proximity testing put her hand less than six inches from the end of the muzzle when the shot to her hand was fired.

When I got home that night, I waited for my daughter to get home from her friend's house and gave her a long hug without saying anything. She gave me a look like, "What's up?" but I just walked away, I could not find any words for the moment. I could handle this work as long as it did not involve children. She appreciated the hug; I needed it. I guess that as a guy, we don't talk much about what is ailing us; we just toughen up and move on. I hoped that I could keep that fiction working a little longer. My wife was a mystic at reading my nonverbal persona. She seldom said anything; she just moved closer and waited to see if I wanted to talk about it. Sometimes I did; other times I could not.

As the evidence was stacking up against Diane Downs, some of the detectives and my overworked associates were talking about this being a slam dunk, but not so fast. In my growth as a forensic scientist for the past five years, I saw the use of scientific method as a working day-to-day model become well developed. The common approach used by researchers was to follow scientific methods to prove a hypothesis or answer a question. In our line of work, this process is dangerous for some of the obvious questions such as, Did she do it? What ultimately happens is that you take the witness stand prepared for the prosecution's questions but are totally unprepared for any alternatives that might be cast forward by the defense.

In application of elimination theory applied to scientific method, it is what can be excluded, not included, that is important. With this approach, when you take the witness stand, you have examined the other possibilities other than your final conclusion and will be prepared for cross examination. On occasion in routine casework, you find there are other explanations for an action other than the one set forth by the prosecution; that can make things interesting.

So, as a forensic scientist, it is important not to seek only information that supports the conviction of a person but to seek elimination of everyone else. We had to rule out the bushy haired stranger, not seek more evidence on Downs. I made this argument several times, as this approach is time-consuming and not often the procedure of choice. I saw it as no other option; if we could eliminate all other possibilities, then she would remain as the only possibility until new information changed our opinion.

As discussed previously in Chapter 5, there is often a fundamental difference in thinking between scientists and police.

A police investigation usually followed deductive reasoning. You add up all the facts that lead you toward a conclusion. You pick up leads until you have probable cause for an arrest: 1 + 1 + 1 = 3. As a scientist, you evaluate all the leads and eliminate what you can, and what remains is the conclusion or conclusions: 3 − 2 = 1. This process may not have been used by all scientists, but this is how I proceed. Finding the toolmark match between the casings in the vehicle and cartridges in the magazine of her own rifle went a long way in eliminating other suspects.

When I started the clothing examination from the boy in the back seat, I knew he had been shot one time in the back. The circle of blood around the bullet hole was dense, but it had not spread more than a few inches. I surmised that the child did not move much after he was shot and was in a facedown position perhaps, as there was no blood on the seat in his location. I was looking for gunpowder, and I could not see a thing through the blood. The common technique here was to wash out the blood and reexamine the area afterward. There is risk here, as one may wash away evidence as well.

Not wanting to alter the shirt by washing, I called Dr. Wilson, the ME, and asked about x-raying the clothing.

We had never done anything like this before, but it would be non-altering for the evidence. We would have to set the energy really low for something like this, but it was possible. He thought about it and said he had a "soft" x-ray machine in the lab that they used for imaging excised breasts to locate cancer. This sounded like what I was looking for, so I took the shirt over to the hospital. We did the x-rays with this machine.

It worked out beautifully, and I could see the outline of the shirt, the blood, bullet hole, and surrounding metal fragments as shown in Figure 17. The metal fragments appeared as white dots. With the aid of the x-ray, I could tease the fabric under the microscope and pull out the trace evidence. Besides gunpowder, the x-ray showed me metal fragments that were new to us. There was no reference to this in the literature, and I suspect no one else had ever tried this x-ray technique to look for them. The tiny fragments were probably lead and were shaved off the bullet by the lands and grooves in the barrel. At least, that was the theory at the time, as we had no literature references to rely upon. Firearms experts are well aware that leading of the lands and grooves in a barrel often takes place. It could be that the new bullet traveling down the barrel pushes lead fragments out ahead of it. This is speculative and new research would be required to eliminate some of the possibilities.

The shirt was a polyester fabric, and when polyester fibers are subjected to intense heat, they swell up. Under the microscope, I could see enlarged bulbous fiber ends around the edge of the bullet hole. We knew this occurred in shootings but did not know how close the muzzle had to be to get this to happen.

I went to the Goodwill store, purchased some shirts of similar fabric, and did a series of shooting experiments in which the muzzle-to-target distance was increased with each subsequent shot. If you ever shoot a weapon at night, and especially if you try to photograph it, you will see a flame exit the barrel after the bullet. This is truly a flame of burning gunpowder, and it travels a finite distance based on the barrel length, powder type, and the amount of powder.

Figure 17

From the heat of the gunshot I could get the fiber swelling out a maximum distance of nine inches in the test shots from a Ruger pistol. This was consistent with distances determined from the stippling pattern on the skin and gunpowder pattern on the shirt as well. If the shooter (bushy-haired stranger) was in the driver's door, this was possible, and he could not be eliminated.

Next was the clothing of the girl who had died in the front seat. The autopsy examination of this child in combination with trace evidence on her shirt provided some interesting information. As mentioned before, she had been shot twice but on opposite sides of the body. The shot through the right shoulder exited the front chest and would not be immediately fatal. This could have allowed for some movement. The distance between the muzzle and the skin was six to nine inches. Perhaps this was the bullet found in the vehicle that struck the dash? The other entrance wound was in the left side. It did not exit and was determined to be a contact shot. I have seen cases in which a high-powered rifle bullet did not exit when the victim was up against a solid surface. Was the bloodspatter seen on the outside rocker panel from this shot? If so, this was not possible from standing at the driver's door. I worked up the proximity testing on all the clothing, made correlations to any skin observations, and prepared a chart of the findings.

In time, I had accumulated and examined about five hundred items of evidence. If the department had not released me from my other duties, I would never have had the time to examine every item properly.

As the time passed, Downs was still under investigation and was eventually arrested for murder and attempted-murder charges. She was not happy that the police were not looking for the bushy-haired stranger and were continuing to focus on her. The children were placed in protective custody early in the case, and she was not allowed to have contact with them, either. I was told that the boy was too young to be a reliable witness, and I was not told what the surviving girl had said, if anything. The proof of this case would rest heavily on forensic science. I would often have one of my fellow scientists review my work to ensure I was correct—what we call *peer review*. Important evidence, such as the toolmarks exams, were reviewed by all the examiners in the lab as we discussed the case and searched for any other explanations.

Diane's clothing was also submitted to the lab as well as a blue-gray postal service sweater from the vehicle that she may have worn in her work for the post office. I knew some folks who worked for the post office and asked if they had any extra sweaters I could borrow for this case. I quickly had more sweaters than I could use. The community was beginning to come to grips with the fact that Diane Downs may have committed this crime. Attitudes were changing.

There were two holes in the sweater that I removed from the vehicle. Both holes were chemically positive for gunpowder. We use two tests that look for lead and nitrates that also give us a size to the pattern of chemical deposition. A close shot will have a very small diameter, and a greater distance will have a larger pattern. The maximum distance for the deposition of GSR is generally in the two-foot range.

The hole on the left was consistent with a contact shot, and the hole on the right was consistent with six to nine inches. It was theorized that the girl in the front seat had the adult-sized sweater on at the time she was shot. The holes in the sweater could be adjusted to overlay the holes in the girl's shirt, so this clothing-overlay possibility could not be excluded. She may have had it on or just pulled over her for warmth. The question was, how to best demonstrate the bullet trajectories to a jury as displayed on the evidence?

I knew a person who made biologically correct dolls for law enforcement for use in child-abuse cases. I asked her to make to-scale dolls for us. She said she could, and they were delivered to the lab a few weeks later. This was fantastic! I was able to mark them with the bullet-trajectory information, put similar clothing on them, and mark that as well (Figure 18). After I had them dressed, I took them over to the hospital to meet the staff again and determine if this was comparable to what they saw on the night of the incident. This was several months since the incident, and it was a good refresher to help everyone get on the same page as well as be accurate.

Figure 18

The next items for examination were Ms. Downs's clothing. Diane was wearing a long-sleeved blouse at the time of the shooting. The testing of the entry hole area of the *left* arm showed a muzzle-to-arm distance of nine inches or less. This could have been a self-inflicted wound, or she could have been shot by the bushy-haired stranger. Neither scenario could be eliminated on this evidence.

Upon microscopic examination of the *right* sleeve, I found several spots of blood less than one millimeter in diameter near the cuff and especially on the side of the shirt near the button.

I had attended a special school in bloodstain pattern analysis a few years prior given in New York state and had done some of my own research in bloodstain pattern analysis. As I mentioned before, I was aware that when a bullet passes through a bloody object, blood will follow the bullet and be called *forward spatter*. If the gases exiting the barrel from a near contact shot enter the body and cause tissue expansion and contraction, bloodspatter will be created as blood exits with the escaping gases back toward the shooter. This is called *back spatter* and does not travel more than a few feet based primarily on the small size of the droplets encountering air resistance. The more objects you put in the way, such as clothing, the less back spatter you will see. What I was seeing was possible back spatter on the end of her right shirt sleeve. Her own injury was to her left arm.

I knew I had something important here and discussed it with the staff. We knew theoretically that this was back spatter, but there was no known documentation in the literature using this pattern to identify a shooter.

After making several calls to experts around the nation, it was apparent that no one else had documented back spatter on a shooter's sleeve. This was not to be taken lightly; if true, it would put the gun in Downs's hand when she fired a weapon in a near contact or contact shot.

We are talking muzzle to skin. But could we eliminate the right hand just being close at the time she was shot? I talked with the others in the lab, and we decided to do some additional work with this. I purchased similar clothing and set up shooting experiments to document the requirements to get such an appearance of spatter. My boss, Chuck, and I dedicated weeks to this testing and looked at several different scenarios. Again, what could we safely exclude, and what choices remained?

The tests confirmed that a contact or near-contact shot with a .22 was required to get the back spatter. The spatter on the sleeve was on the wrong side to have the right arm close to the left when she was shot. Our testing also showed that the weapon and the fingers often block the deposition of spatter on the sleeve of the hand holding the weapon except for the area of the sleeve where the button is located. The hand is fairly straight here and allows for the passage and deposition of the spatter on the sleeve.

The absence of any of this in the literature was a concern to me. My gut feeling put this all together with the shot into the side of the girl outside the vehicle, but being too obvious was worrisome. The muzzle-to-sleeve distance for Ms. Downs's own shot to her left hand was too great. This got out to the DA, and I was on the hot seat. The implications were too much to pass up. He wanted a report describing our research and a conclusion that Ms. Downs held the firearm when the girl in the front seat was shot. The information about the case implied this was when the spatter was deposited, but the bloodstains were too small to be blood typed.

I was hesitant. The division director came down from Salem as he had been advised of the situation in this high profile case and wanted to talk. He was a major but was also a scientist, and I normally trusted his judgment. He looked at the work we had done and the blouse and concluded there could be no other choice than that it was back spatter from the shooting of the girl in the front seat. My immediate boss was also in concurrence. I was sweating. From the lecture I got, you would think I had done nothing right in my life, and what was I thinking? My only out was the major's own words. When I first came into the lab system, the division director had two rules that everyone was to abide by: *You did not testify to anything that you were uncomfortable saying, and you did not run any tests that you felt unqualified to perform.*

These were not all the rules but the personal expectations for your performance with the Oregon State Police. These were words to live by when the lives of others were in your hands. I was concerned in that this was new science. We had not published on this yet, and it had not withstood the test of peer review by other scientists. I punted. He and my immediate boss were pissed and said that if I was not going to do it, they would. I said, "Go for it," in an uncertain, quiet tone of voice. We all served at the whim of the superintendent. We could be fired without just cause if he deemed it so. A call to him about a violation of a direct order, and I would be history. When you like your job and also have a family to feed, this is not an idle consideration. I'm not sure what happened next or how this went over with the DA, but my bosses did not write a report or testify. I suspect both of them knew I was right, despite the pressure they were under for me to perform.

But word got out on what I had, and just before trial, the defense attorney came to see me. Oregon law has dual disclosure rules, and the defense is entitled to interview the witnesses. I showed him the evidence, let him look through the microscope, and told him what I thought. His only question was, what was I going to say on the witness stand? My response was that I couldn't exclude that possibility, but it had not been adequately researched or peer reviewed.

The sign of a professional in this business is that you don't know what you don't know and tell yourself that every day. This is a step up from the earlier years in your career when you think you know everything. We continued to research this back spatter and a year later, after trial had concluded, we published our findings in the *Journal of Forensic Sciences*.

It was known early on that Diane Downs had possessed, at one time or another, a .22 weapon. It was stated that it was a Ruger semiautomatic pistol. I had heard that there was a zippered case, the weapon, and two magazines. This weapon was never recovered. What could have happened to the weapon? Are there characteristics on available evidence that could identify this weapon? What could the manufacturer tell us? Were there any other makes and models of weapons that could match the characteristics seen on the evidence? What about the bullets that were recovered? Any other evidence out there that would be helpful? If there were any other makes and models of weapons possible, they had to be included in the discussion. Early in the case, based on information provided by the sheriff's office, we purchased a Ruger semiautomatic pistol for our testing.

During the investigation, the sheriff's office (SO) learned that Diane Downs had fired the Ruger on several occasions when she lived in Arizona. One time was during an argument; a shot was fired, and the bullet allegedly went through the floor of a mobile home. A team went down there to see if they could find the bullet. They did locate a hole in the bathroom floor.

There was no way to access the soil under the mobile home for the bullet, so they moved the mobile home. Not a small endeavor. A lead bullet was excavated in an area that would have been under the bathroom, and it was rushed up to our lab.

It was .22 caliber lead typical of a rimfire cartridge, but the soil had destroyed the fine striations used for comparison purposes. I could see an outline of lands and grooves from the barrel, and they were consistent in number and width with a small group of firearms that included a Ruger.

Downs occasionally shot the weapon at a local firing range also, so the guys went to the range and started picking up .22 cartridge cases and putting them in bags. Like a lot of shooting ranges, there were lots of .22 casings on the ground. I received into the lab several bags, each with about 250 .22 cartridge casings. In a normal comparison situation, it could take an examiner all day to reach a conclusion with just one unknown. So how do you handle so many? Among the different manufacturers of .22 rifles and pistols, the shape of the firing pin can vary. The shape of the pin in a Ruger was not unique to a Ruger, but it was enough to use as an elimination tool for many of the casings. Among the hundreds that remained, some were so old that no toolmarks of value remained, and they were tossed as well. For the rest, it was a tedious process, days in the making, and in the end, I did not find one cartridge case that possessed toolmarks that matched the unknowns in this case. This was a big disappointment for the team.

During the pretrial phase, there were a number of hearings I attended. Diane knew who I was. We occasionally made eye contact, but not one word was spoken between us. When we were in trial, she watched me carefully, but I got the feeling she just did not know what to make of this forensic stuff and was unprepared to truly understand what it meant. I could be wrong, but that is how I rationalized it in my mind.

Now to focus on the bullets from the vehicle and the children. We were not going to get a match by microscopic comparison due to their damaged condition. Based on class characteristics, they could have been fired from the same weapon; we just did not have that weapon. But could we discover possible weapons it could have been fired from? The answer is yes. The FBI had created a list based on the number of lands and grooves and their width. The land and groove measurements cross-referenced to direction of twist and provided a list of possible weapons. The list was not inclusive of all firearms available, but it at least gave some guidance. I called them to get an idea of how many weapons they had examined and measured by that time, and the list was over ten thousand. We had six lands and grooves with a right twist. The graphic below demonstrates what you see when you look down the barrel of a weapon and the lines created on a bullet that passes down that barrel. Figure 19 was a transparency used in trial to educate the jury on barrel lands and grooves and their marks left on a bullet.

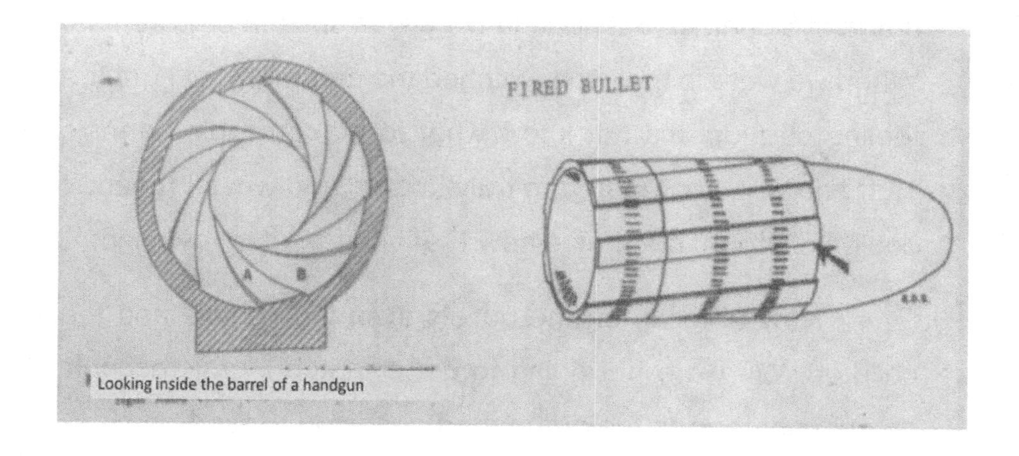

Figure 19

Through the investigation, the team had an idea of the possible serial number of the Ruger in question. I called Ruger and talked to one of the technical reps and requested test fired bullets and cartridge cases from pistols they might have that were close in serial number. I received casings and bullets from several weapons. They also confirmed that the model in question had six lands and grooves with a right twist. The measurements that I had for the unknown bullets were consistent with their manufacturing measurements.

The FBI and the Ruger were in agreement, but the FBI list also had a Browning Nomad as a possible fit.

At this point I called several other examiners in other states looking for ideas, but none was forthcoming beyond what I had already done. The conclusion was the cartridge case toolmarks were consistent with a Ruger but could be from a Browning as well, so I listed the Browning as a potential weapon along with the Ruger in my report.

The question still remained: What had happened to the weapon? The Mohawk River was next to the roadway where the casings were found. So the sheriff's office had divers search the river for the weapon, but none was found. One day I was standing in a store looking at gun cases and suddenly realized that the padding was a soft foam that probably floated. I purchased a case for a pistol, took it back to the lab, placed the Ruger with two magazines in it, and zippered it up. Then I placed the case in a water tank. It floated so high that it did not get the zipper wet. If the weapon was tossed into the river inside the case, it probably floated away.

We were getting closer to trial, and the question was how to present all this forensic stuff to the jury in a manner that they would understand. One common theme with jurors:

If they do not understand, they will most likely disregard. We just could not afford that to happen. I was into posters, and we built about a dozen posters with graphics and photos of the evidence for starters. In conversation with the DA, I decided to generate a notebook of graphics that paralleled my testimony. The jury, judge, and all attorneys would have a copy to follow as I presented the evidence on an overhead projector.

The jury could take this notebook with them in deliberations as well. When I finally testified, I watched the jury study the pages and take notes; it was working.

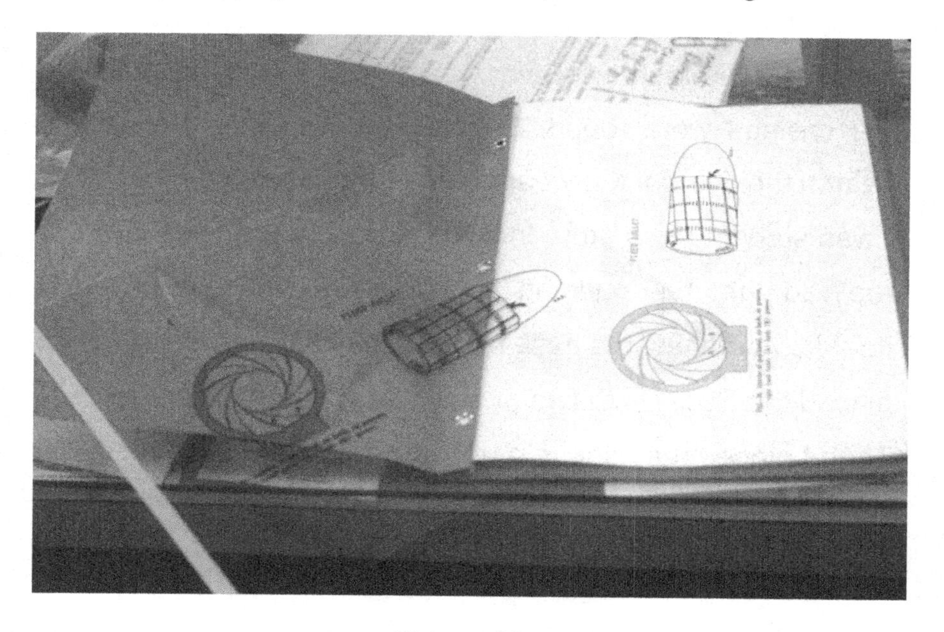

Figure 20

I am not sure if anyone prior to my testimony had developed a notebook for the jury. I guess someone had to be the first. Figure 20 was the notebook and a transparency.

One evening before trial, a group of us met at the Nissan with the dolls, trajectory rods in place, and went over various scenarios on how the incident may have gone down. We put a doll for each child in the appropriate seats and stood back to discuss. As shown on the next page in figure 21, placing the doll representing the child in the front seat was an epiphany for us. The evidence was all coming together.

The second photo, Figure 22, would work for the location of the back spattered blood on the rocker panel, but how the feet were positioned was an unknown. The diameter of the doll was considered close to scale for the child, and I measured the distance from the floor to the muzzle of the pistol: nine inches. In order to get the spatter on the rocker panel, the shot had to be at an angle that would allow the spatter to be unobstructed to the panel. This worked at a distance from the panel of one foot. Go back and look at the photo in Figure 14 constructed on the day of the scene processing. The photos in Figure 21 and 22 were almost a year later.

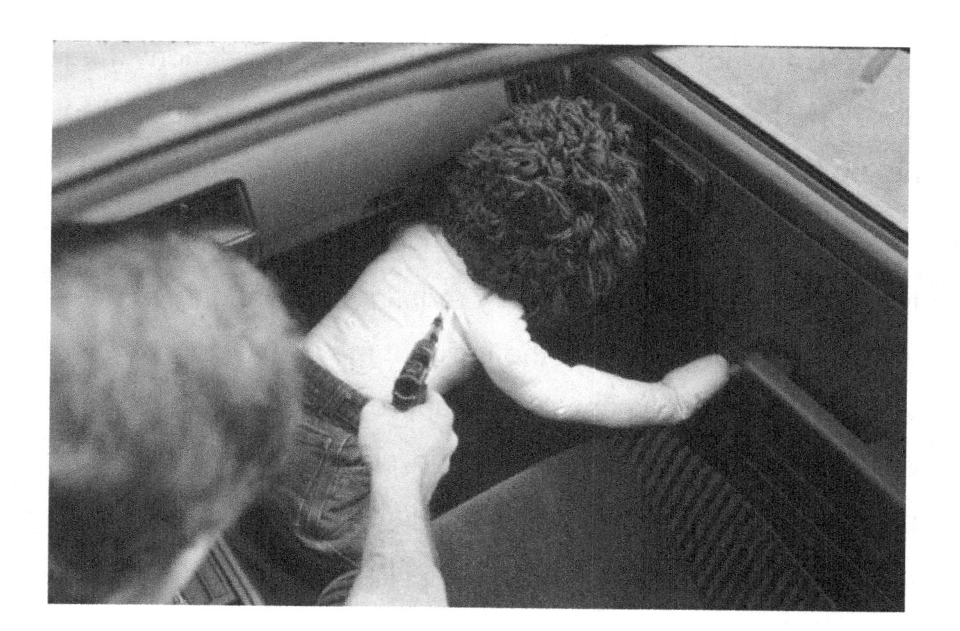

Figure 21

Figure 22

Now the question was how to get this before the eyes of the jury. You couldn't take a car into the courtroom, but you could have the jury travel to the car. But the car had a hardtop, and it was difficult to see inside as a group. Video was considered but not very popular at the time; we were all perplexed on how to display this crucial step. One morning one of the investigators was on his way to work and passed a storefront where a man was building a display for the store. The investigator stopped and talked to him, showed him an advertising brochure of the vehicle, and asked if he could build something like it. He said he could and manufactured a fantastic to-scale model of the vehicle interior complete with working doors, folding seats, and removable hardtop. The model was built for a bargain $800 (Figure 23).

Figure 23

I could place blood and other evidence on the model and place the dolls inside as well. If needed, the model could also be tilted toward the jury for a better view. We were set for trial.

In the last month prior to trial, we were busy with meetings and discussing how to present the evidence. It was important that the jury understand the science if they were going to reach a decision in the prosecution's favor. To improve the presentation, we decided to break up the various scientific areas with different presentation formats.

As mentioned, there was also the notebook. We had a dozen large posters, enlarged photographs in eight-by-ten format by the hundreds, the actual evidence, and the model car with the dolls. It would be quite a show, and I was so nervous; there was little sleep possible.

The DA put together a group of office staff for mock trial. I would present my stuff and wait to see if they grasped the concepts and facts. I got a thumbs-up on some of it and a thumbs-down on other parts. The thumbs-down was on a presentation about bloodspatter and its implications. I went back to the beginning and made new slides in PowerPoint and tried again. We kept this up until the staff was able to understand the implications of the evidence. As I looked back, this would not have been possible if I had not been separated from my other cases just to work on this one.

Suddenly everything became more complex when we heard that Melvin Belli from San Francisco would defend Downs. Mr. Belli was a nationally famous defense attorney. Not long afterward, a person from his office showed up, asked a lot of questions, and left. I don't know if the outcome would have been any different, but we all breathed a sigh of relief when we heard that Belli could not make it for trial. He had other obligations, and the judge would not postpone the trial date. I was told that he had an appointment with the pope and that the date could not be rescheduled.

During the last several months, the case had drawn national attention, and it was on the news regularly. Downs or her attorney were personally on the news regularly, too. To add a little more flavor before trial, we learned that the author Ann Rule was going to sit in on the trial.

The trial date finally arrived, and I was as ready as I would ever be. I kept researching my findings, considering every aspect, just looking for anything I may have missed. I kept second-guessing myself about the spatter on the sleeve but was determined to hold my ground. Just before trial, we heard there were defense experts as well to spice things up. I knew that the surviving girl was going to testify but did not know what she was going to say. If anyone else knew, they did not tell me. She reportedly had been separated from her mother for the last year, and we were not certain how she would respond when she took the stand. It had been a year since I was separated from my other duties, and as far as I knew, I was going to have to prove this case through forensics. She was the only reliable witness, and I had no idea what she was going to say. None of us knew what that moment was going to bring for certain.

The doors to the court room opened at seven in the morning, and no one was admitted before. The seats were on a first-come, first-seated basis.

I always had a seat behind the DA but was outside on some days waiting with everyone else for the doors to open. Two people with the news networks I remember most were Lars Larson and Ann Bradley. We would migrate together and talk until the doors opened about the dynamics of the case in general but never about the evidence. The courtroom was full of the exhibits that we prepared, and it was difficult to get around from one item to another during my testimony. I spent three days on the stand and on the floor using each one. I was anxious to show off the rocker panel from the Nissan. It had been packaged since my first examination, and I had not seen it since.

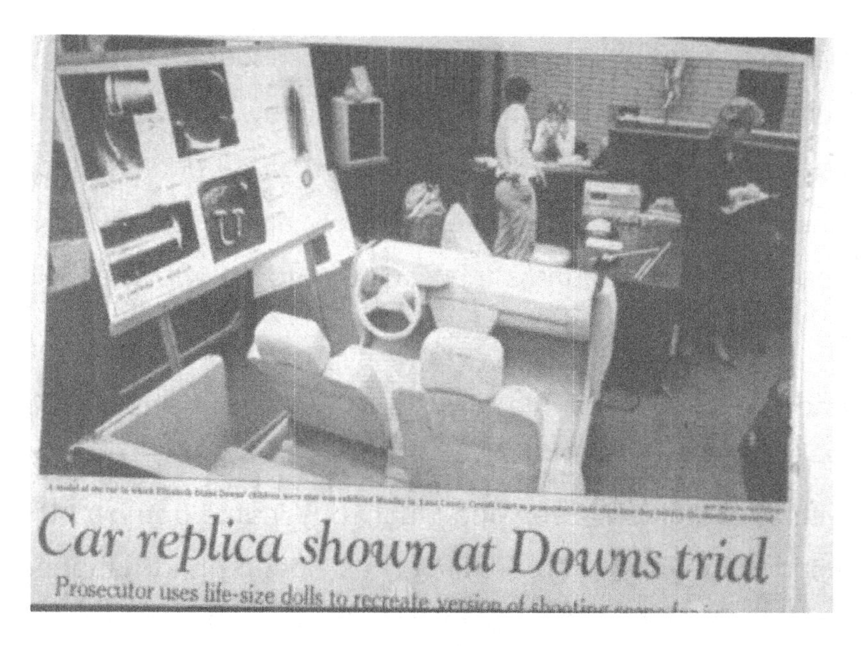

Car replica shown at Downs trial

Prosecutor uses life-size dolls to recreate version of

Figure 24

I did a tutorial for the jury on bloodspatter and now had to show the findings from the vehicle. I carefully unwrapped the long metal object and turned it over, expecting to see the spatter. It was gone! Oh my God! In the past year, it had separated from the metal. As the package was handled, it had probably been jolted a few times, and the blood had loosened and fallen. Anyway, it was no longer present.

It is not often I am speechless, but this was one of those moments. After a pause and without the ability to communicate a plan to the DA, I stated that it was reasonable for this to happen. My heart rate was so high it probably was not measurable at that time. The DA started to hand me photographs of the blood that I had taken prior to removal of the rocker panel, and we moved on from there. I was deeply concerned about how the jury might view this mishap and called for a meeting with the trial team at the end of the day. They said that I had recovered nicely and that they did not see a need to readdress the issue and to just move on.

In my testimony demonstrating positions of the dolls and the position of the shooter for the front seat shots, I stepped into the model car, and my foot went clear through it with a loud crash. Everyone laughed. What to do now? I just withdrew my foot and continued as nothing had happened. My brow was sweating, but what else could I do? Smile and move on.

The cross-examination centered around the blood on the rocker panel of the car. Because of the circular shape of the droplets and their upward and vertical appearance, there was no way the vehicle could have been moving at the time the blood was deposited. Through the elimination process, I had already covered all other possibilities. I was prepared for the cross-examination. The defense attorney offered that perhaps when the girl in the front seat was removed at the hospital, a blood-soaked sleeve from the postal service sweater over her had fallen free and swung downward, depositing the blood on the rocker panel.

Alternately perhaps the nurse stepped in blood outside the vehicle as the girl was removed and spattered the blood on the rocker panel.

Without hesitation, I said no, I did not believe either scenario was possible. The direction of the swing was wrong, and I had never seen stomped blood get that high up a wall. But as many times as I had gone over this evidence, to my surprise, I was not immediately prepared for those questions. As soon as court was over for the day, I went back to the lab. I found another victim (friend) to draw blood from, and I was dipping sweater sleeves in blood and swinging them back and forth in as many ways as I could think of. As I suspected, it would not produce the results visible on the vehicle. The rocker panel curved under the body of the car below the door, and the surface was not available from directly above or from inside the vehicle.

Besides, with the sweater, it was hard to shake any blood loose in the size of the droplets observed. You had to snap the sleeve like you would a whip to get any small spatter. This was not how a nurse would handle an injured child.

After I had completed this project, I got more blood from one of my friends and went to the hospital parking lot. The parking lot of the hospital near the ER was uniform asphalt throughout. I picked a location near the entrance, and I set up a cardboard circular wall. I poured the blood in the center of the cardboard circle and then stomped in it. At first there was no spatter at all, but subsequent attempts created a few spatters but only a few inches above the pavement. None reached the height of the spatter on the rocker panel. The reason for this was simple. Pavement is not smooth. It is a surface with several peaks and valleys. The blood quickly settled in the valleys, and when you stomped on it, all you got was pressure on the peaks; the majority of the actual blood was not impacted.

It was late when I created more posters of the sweater tests for court next day. The defense expert was on the stand first and under cross-examination by the prosecutor. He was asked to use the mockup vehicle, a doll, and sweater to demonstrate how a child would be removed from the vehicle in an emergency.

In several attempts, he could not get the sweater sleeve to swing in a manner that would produce the spatter seen on the rocker panel.

I retook the stand and explained what I had done using the new posters. Ultimately the defense expert agreed with me.

Cross-examination on the firearms stuff was short. The matched extractor toolmarks were conclusive. I suspect that the defense experts had informed the attorney that I was right and no one was going to object to my conclusions. Sometimes the best defense is to not dwell on the indefensible.

Finally, I was excused, and the trial moved on to other phases. I felt good about what I had done and had made frequent eye contact with several of the jurors. I think they got it. Diane would sit in her seat and take notes. She would look at me as I spoke but was careful not to engage me through her expressions. She was well coached by her attorney.

As it turned out, the surviving girl, who had not seen her mother in a year, testified that her mom had fired the shots. The pressure was off somewhat now. My testimony moved to be support for her testimony. After a year of intense investigation and many sleepless nights, I got it right, as confirmed by Diane's daughter. Diane Downs was found guilty on all counts.

Chapter 13 Rural Oregon and an Officer Murdered

In our lab system, we had three major labs down the I5 corridor and four smaller labs strategically located in other areas of Oregon. At times the lab director would go on leave, and someone would have to fill in. During one of these leaves, I was chosen to go to the Pendleton lab to run it for a couple of weeks. Jon, the director, was a unique fellow. He operated on a very small budget with just his secretary in the lab. When I arrived, only the secretary was working. I introduced myself and started picking up case files for testing. I noticed a number of test tubes in the dish rack and realized that Jon washed his test tubes! These were disposable test tubes and were to be thrown away after use. But not Jon; he washed them.

The first day I picked out a case that required a revolver to be test-fired, thinking this should be simple enough. I found the ammunition in his supplies, picked up the weapon, and asked the secretary where the bullet tank was located. The secretary laughed and said it was in the freezer. Say what?

Upon opening the freezer compartment of the refrigerator, I found a coffee can containing ice.

She said, "What you do is place it on the floor and shoot into it." I was beside myself. I was holding a .38 Special revolver, and she wanted me to shoot into six inches of ice? In a can on the floor? Thinking about this, I circled the can several times and eventually said to myself, "What the hell," and cocked the hammer and fired into the can. No protection for lead exposure, no protection if I missed, and much explaining if I did. The revolver went off, and the bullet penetrated most of the ice. Now all I had to do was wait for the ice to melt and recover my bullet. Who would be so ingenious to think of this? Naturally it would be Jon.

I needed about five test-fired bullets, so the process took a while to refreeze the water. But to my surprise, the bullets were in good shape for comparison testing. In the past, I have always used a large water tank especially designed for bullet recovery. This was all new to me. While I was working, I met several of the local law enforcement officers, and a few years later, this proved to be a valuable association. I got a lot done while I was there, but Jon did have to order some new "disposable" test tubes when he returned. I disposed of his current supply.

Years later Jon did a presentation at a forensic meeting on a homicide that occurred in a rural wayside in eastern Oregon. This case exemplified Jon's creativity within a limited budget and still capable of providing an outstanding service. If you watch *CSI* on television and see all the latest tools available, think about what I am about to tell you.

A young woman had witnessed a drug deal and was strangled by the bad guys. Then her body was transported in a sleeping bag to a remote area for disposal and dumped into the hole under an outhouse at a rural campground. No one frequented the campground except during hunting season, which was months away. The body went undiscovered for 18 years. At some point, a witness advised the local law enforcement where the body was located and Jon responded to the scene.

The outhouse was moved, and after a preliminary dig, it was apparent that there were only skeletal remains below mixed with other "outhouse" contents. Most of the body was still in the sleeping bag and was taken to the medical examiner's office for autopsy. Jon shoveled out the other contents into large plastic bags, and took them home. He had no place at the lab to process this kind of "outdoor" evidence, so he did it in his backyard. He built a set of sifting screens from material picked up at a local hardware store and hooked up the garden hose. Then he began going through the contents of the plastic bags with the screens on sawhorses. He found additional bones, cleaned them up, and sent them on to the medical examiner's office. The dental records for this woman had been destroyed long ago and the body could not be identified with certainty. Jon went back to the scene again and collected more material for examination. The only identification now possible for the body was a necklace and ear rings that she had worn that her mother had given her. Jon was successful in finding both items using his sifters and a garden hose. Mom confirmed these were her daughter's jewelry, thus an identification was made.

I cannot imagine anyone doing this kind of back yard work today, but Jon was not above doing what it took to get the job done with the resources he had available. I worked with some extraordinary people in those early years, and Jon was one of them. Off duty he would dress in old jeans with suspenders and an old white T-shirt, and just looked like one of the locals. He was somewhat reclusive, lived in a rural area and had a pet python to keep the neighbors away. I suspect that year the snake found the backyard grass to be a little greener than in the past.

In the late 1980s, Jon had retired, and a new person was running the lab who had limited crime scene experience. A police officer was murdered in the small town of John Day, and the lab needed help. The local law enforcement agency contacted the superintendent of state police. The superintendent contacted the lab and told my boss to send me, despite the fact that I was not on call.

As ordered, I grabbed my call-out stuff and headed out. It was a long drive, and I had not been in this area of the state in many years and was uncertain of the shortest route. I did not have a GPS to guide me.

When I reached the small town of Sisters, about halfway, I saw a sheriff's patrol car beside the road and stopped to get directions. I started to say who I was, and he interrupted me. He said he knew who I was and said that the law enforcement community on this side of the state has been waiting for me. He gave me directions and wished me God's speed in getting there. I was beside myself; I had no idea how important my presence was to this case and how my role had changed from assisting to directing the forensic investigation.

When I got to John Day, I met with the other crime lab guy, Steve. We were to meet with a couple of local law enforcement officers, or so I thought. When we got to the meeting, there was a room full of representatives from many agencies waiting for me. I was beginning to feel uncomfortable and fearful that I could not meet their expectations. They had lost one of their own and had the suspect in custody. It was expected that I was to find the evidence to hang him high. These guys were well intentioned but serious.

The death of one of their own in this small western town might have required only a tree and a short rope not many years before.

High-profile cases do not always provide evidence laden crime scenes, and success is not guaranteed. We got an overview of what they suspected had happened at the scene, and we were on our way.

The deceased was a reserve officer, and he had received a call from dispatch to respond to a domestic assault where a boyfriend was beating up his girlfriend. The department was so strapped for funds that the officer did not have his own equipment and had to borrow a weapon from the department locker to respond.

This was exactly the kind of situation Pierce Brooks documents in his book *Officer Down, Code Three*. The reserve officer entered the residence alone and contacted the boyfriend in the kitchen. Entering a domestic violence scene alone is never a good idea. A fight quickly ensued, where the boyfriend picked up a block of firewood and struck the officer several times. The officer attempted to retreat but died on the floor near the front door from head injuries without firing a shot.

We traveled to the house and did an overview of the scene first. We photographed extensively, and in the search, we discovered several bloodstain patterns between the kitchen and the front door. Bloodstain pattern analysis cannot always tell you the sequence of events, but it can assist and support the statements of others, such as the girlfriend.

She said that the direction of travel for the officer during the assault was from the kitchen through the living room toward the front door as he tried to escape his aggressor.

In the analysis of bloodstains, it takes at least one blow to bring blood to the surface of the skin to be available for a second blow that would create a bloodspatter pattern on adjacent objects. I found several bloodspatter patterns on the ceiling, the walls, and the floor. Making sense of this seemed overwhelming at first. At first we separated the bloodstains into patterns and numbered them, and then we measured the length and width of several droplets in each so that we could triangulate the two-dimensional and three-dimensional location of the officer each time he was struck.

At the entry from the kitchen to the living room, he was standing, and the center of the pattern was about five feet above the floor. A few feet away, he was on the floor, and then he was standing again, based on the pattern being close to the ceiling. As the assault on the officer approached the front door, the patterns were close to the floor, and the officer was found dead a few feet beyond just inside the door. A large spatter pattern was visible here, where it appeared that he was struck several times. All in all, I had about six different positions at the time the officer was struck, excluding any strikes somewhere near the kitchen that had started the blood flow.

After I returned home, I wrote a long report on the crime scene, went about my other work, and did not hear any more about this case for several months.

The boyfriend eventually pled guilty, and I heard that during sentencing, the judge read directly from my report to the accused. The judge would describe a position based on bloodspatter pattern and state that the accused could have stopped at that point. Then he read another position in the sequence and stated that he could have stopped at that point. I was told that the judge's voice rose each time he read a new position, ending at the door. The assailant received a life sentence from the judge.

After the sentencing, the superintendent called me and thanked me personally for the work I did, which added to an already good feeling about the success of this case. I never forgot how well I was treated by local law enforcement while on scene.

Twenty years later and after I retired, I was in John Day and in a restaurant having lunch when a guy came up to me and asked if I was Jim Pex. I said yes, and he said, "Let me pick up your check for you."

I did not know who he was and inquired about this sudden interest. He said he had been a police officer when I had come to town many years ago and processed the scene where the officer had been killed.

He said anyone in law enforcement for a hundred miles knew who I was, based on what the judge did. The judge reading my report was the highlight of the case, and the community was forever thankful for the work I had done. I had never heard any of this.

He wanted to show his personal gratitude by picking up my lunch. I was nearly drawn to tears. Sometimes you just don't know how you impact others in doing your work that sometimes seems ordinary to you. I have never forgotten that experience. I made a difference that day to that community.

Chapter 14 The Alibi Tavern Case

Mike, one of my coworkers, had worked a homicide scene at a tavern in the early 1980s. On this evening, it was reported that the tavern owner was down to only a few patrons by closing time, at 2:00 a.m. Sometime around closing, he was beaten with pool cues. He was struck so hard that several cues were broken and lying on the floor of the tavern. He was also stabbed with a pair of scissors and subsequently dragged out of sight into a back room. There were large pools of blood on the floor in the main area of the tavern, and some shoe impressions (figure 25) could be seen in the blood. The next day an individual was caught with the owner's vehicle and transported to the police station for questioning. Mike had finished the processing of the scene by then and went to the police station to check in and take hair standards from the suspect.

While he was collecting hair samples from the suspect, he noticed blood on the suspect's clothing. He alerted the detectives, and the suspect's clothing was quickly seized for examination later. A quick look at the shoes revealed that the sole pattern was consistent with what Mike had seen at the crime scene in blood.

Figure 25

The suspect had to have been at the crime scene while the blood was still liquid in order to create the pattern. No one considered how long that might be, but it was not important at the time. The suspect was arrested for murder and at trial was found guilty and sentenced to a lengthy stay in the penitentiary.

Apparently at some point during the trial, the weather was bad, and a deputy was sent to pick up a juror who was unable to get to the courthouse on her own. During the transport, the deputy and the juror discussed the case including facts not in evidence. The individual's conviction was appealed, and the case, as sometimes happens, was remanded back for retrial a few years later.

I had done some blood typing work in the case, so Mike and I traveled to the DA's office for a pretrial meeting to discuss the case. The DA was quite concerned. The defense had a new wrinkle to their story. He told us that the defense was going to say that his client was sitting in the bar when some suspicious looking guys came in. The subject had just been released from prison and did not want to hang around for fear something bad was going to happen. He said he left, walked the streets for at least a half hour, and then returned to the bar. When he walked in, there was no one visible inside the bar. He looked around and found the owner dead in a closet. The accused needed money, so he opened the till of the cash register and took all the money, found the keys to the car, and drove away. He was guilty of robbery and auto theft but not murder.

There was an alibi for the murder at the Alibi Tavern.

The blood on his pants and his shoe impressions were from being in the scene after the owner was already dead. There was no mention as to how the blood had got as high as the top of his pants, but Mike and I were sure we would hear a theory at some point. The DA did not know how he was going to get around this time-line issue. Mike and I talked about it with the DA and did not have an answer, either.

While we were in his office, I studied the pictures from the scene as we talked and had an idea. I said, "What if we could put the accused there at the scene at the time the blood from the owner hit the floor?"

Mike looked at me like I was nuts, but the DA said that if we could do that, he could get a conviction. The individual bloodstains disturbed by the shoes looked like the action had occurred while the blood was really fresh and had no time to dry. I just did not know how long that would be. Mike and I drove back to the lab discussing the case and blood drying. We were not aware of any earlier studies that looked at the drying time of blood, especially in a circumstance where this would be the prima facie evidence for a conviction.

Several questions emerged during our discussion: How long does it take for a droplet of blood to dry? What were the interior environmental factors at the tavern at the time of the incident? Did it matter? Does the size of the droplet matter? What are the physics principles involved? We had some work to do. As this appeared to be new science, we also had to be very careful. Both Mike and I were trained in bloodstain pattern analysis. Both of us were convinced that the drying time concept was possible, but how to start? A quick literature search, and nothing was found that would be helpful in this case. We also had to know how long after the blood hit the floor did the shoe disturb the blood.

The shape of a single bloodstain is common knowledge. When on a flat nonabsorbent surface, the liquid is thin at the edges and thickest in the center, as seen in figure 26. The rate of evaporation of water from a liquid is equal across the top of the droplet on a nonabsorbent surface. Therefore, the blood at the edges dries first, and the drying continues until the center is finally dry; the concept being that as the droplet dries, the thin outside edge is replaced with another thin edge with a smaller circumference. This process continues inward until the entire bloodstain is dry, as seen below, in minutes under standard indoor conditions. All of this is based on Henry's law of physics.

Remember when you were in high school or first year physics in college and there was a chapter on the Ideal Gas Laws? Something you memorized for the next exam and never heard of again? Henry's Law was one of them. A quick explanation of Henry's Law was that a liquid will surrender a portion of its content to the atmosphere as a gas directly above the liquid until the atmosphere reaches an equilibrium with the liquid at a given temperature. This happens in a closed container, and the major variable is temperature. If the liquid is not in a closed container, the moisture (gas) above the liquid disappears into the atmosphere, and the liquid keeps giving up more of its content.

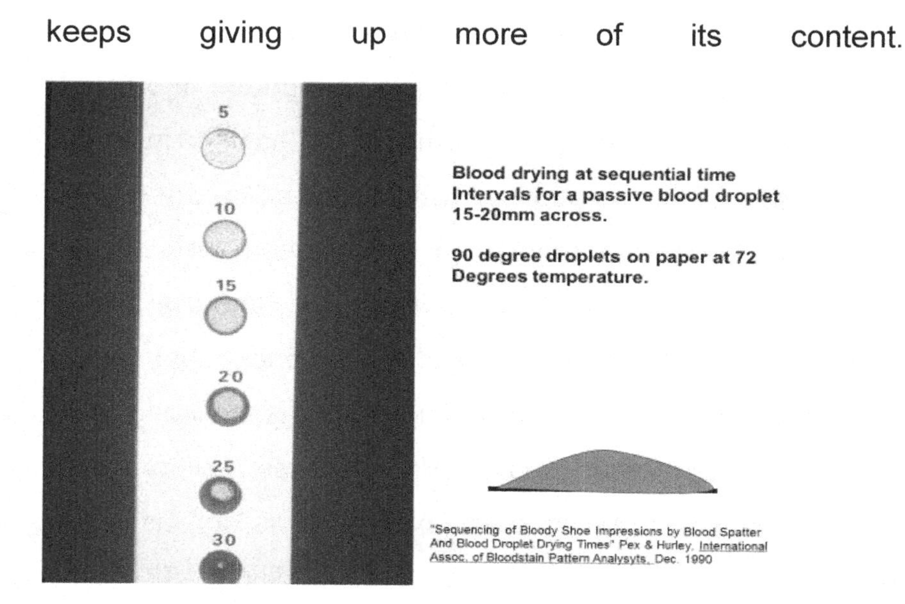

Figure 26

Eventually all the liquid will disappear; this is evaporation. In the case of blood, the primary liquid lost to the air is water. The chart in Figure 26 shows the progression of blood drying.

An easy explanation of the effect of temperature is heating a cup of soup. When you first put the cold soup on the stove, you cannot smell it. But when heat is applied, you can, because the heat allows for more of the organics and water in the soup to enter the atmosphere above the soup.

If you leave the soup on the stove too long, it will all evaporate away. We needed to know what the temperature was in the tavern.

We also did our experiments with anticoagulated blood. Do anticoagulants used in our test tubes make a difference compared to fresh blood? Does humidity have a factor in this? How do you go back and measure humidity? Will it be the same now as it was when the crime was committed? Does anything on the floor make a difference? Mike and I were faced with these and other questions when we began testing blood inside the lab and measuring how long it took for it to dry.

Fortunately, inside commercial buildings, the heat and humidity are both controlled. This included the tavern, and the records were available. The contractor who maintained the heating system of the tavern had records that went back far beyond when the incident occurred. We had that base covered, and to our surprise, the environmental conditions were still the same many years later.

We went back to the Alibi Tavern before they opened one evening and set up for testing. The outside showed a lot of wear with old yellow siding and a brown roof from the time of the incident, and the inside was not much better. It was cleaner than it had been in the past and still had an old-time cash register, pool tables, and an open space for dancing.

With the current owner's permission, we placed several drops of blood on a floor tile, started a stopwatch, and wiped away each drop at five-minute intervals. This was going to be as close to the incident conditions as we could get.

Not only did we see a reproducible pattern versus time, but we also saw that smaller drops dried faster than larger ones, an additional bit of information. This made sense; the water content in a smaller drop was less. In subsequent testing, I found that droplets down to about a millimeter in size will dry in seconds after they strike a surface. Sometimes the shapes were distorted, and the tiny droplets looked like small volcanos, others looked like a life ring.

The research looking at humidity with a constant temperature was more complex. First, we needed an accurate method of measuring humidity. After making several phone calls, we were able to purchase a swinging cyclometer. Simply described, it is two thermometers hooked together with a wet cotton cloth on the tip of one. The cotton tip was wetted, and the instrument swung around like you were going to throw a rock in a sling. Then the values on the thermometers were read, and a calculation was done. This provided the humidity.
Prior to this study, we had never seen one of these devices.

Since indoor commercial buildings were strictly controlled, we went down to a local swimming pool to get the humidity there. As expected, it was high, over 80 percent, so we got out our blood samples and started dropping blood on a tile while young children were swimming in the pool. Our testing created quite a sensation, and it was difficult to continue the experiment with children asking lots of questions while they looked over our shoulders. When we told them that we were police officers, they were even more excited. In conclusion, what we found was that when the humidity was less than 100 percent, evaporation would take place. There is still room in the atmosphere for more water.

Outdoors there is the dew point. This is when the humidity reaches 100 percent when the temperature drops, creating a saturated environment. Hence droplets of water form on the grass, and we have the morning dew. If the temperature was less than the dew point, the times for evaporation of the bloodstains were measurable and reproducible.

Since we had our own caseload to keep up, it took about six months for us to conclude that we had a solution to the crime scene time line. We reported this to the DA, and he discovered the information on to the defense. We could not find a literature reference on blood drying times applicable to our situation, so we wrote up what we had and published it in the International Association of Bloodstain Pattern Analysts

Newsletter. Again, Mike and I were knocking on the door of new science.

After looking at our findings, the defense hired their own expert. It was a physician who was a hematologist, a specialist in blood or blood diseases. He did not agree with us, of course, and felt that the drying times we specified did not consider several factors.

We did not have the physician's credentials, and the defense held the trump card for the moment. I suggested we have a conference call with this guy and see what he had to say. We had the DA set up that conference call, and the doctor said that there could be no standardization of blood drying time and that coagulation had to be considered as a factor. I thanked him for his opinion and advised him that I, too, was familiar with coagulation stages and reminded him of the multiple steps involved in intrinsic and extrinsic coagulation.

I asked him which one he thought was responsible for this concept of his—intrinsic or extrinsic. Intrinsic coagulation takes place via factors found in the blood. Extrinsic coagulation takes place based on factors found in tissue. When tissue is cut, these factors are released. He did not know. He said that in surgery, he had seen blood fall to the floor and believed some coagulation occurred. More important than the coagulation issue, I wanted him to know that he was not the only one who was knowledgeable on blood.

"Did you have proof of that?" I asked. "Did you get down on the floor with a probe and look for fibrin?"

The answer was no. I reminded him also that laboratory coagulation tests were conducted at body temperature. Did he know the temperature at the floor? Did he know if this made any difference? Did he remember from his biochemistry that the speed or even the action of some enzymes responsible for coagulation may be influenced by temperature? Regarding the enzymes, he answered yes. I asked several other questions that he could not answer. All were based on blood on the floor, not in a test tube. I think it was a surprise to both the defense and this physician that someone else would know this stuff, especially someone working for a police department. We must have been successful in discouraging him; we never heard from him again. My ten years in a clinical laboratory paid off again.

Mike and I went back to the original observations and prepared a report for trial. I also went back and repeated testing with fresh and anticoagulated blood to show there was no difference and coagulation was not a factor. By the time we finished our report, a year had passed since we had started this research.

At trial, Mike went back over the original evidence he had recovered from the scene, such as the shoe impressions. It takes time to introduce new science to the courtroom and to convince the judge to let it in, so I was on the stand quite a while explaining what we had been doing for the past year.

By law, the benefit of any doubt must favor the defendant. I pointed out in the photographs where some bloodstains were so fresh that they did not have the time to form any ring before they were disturbed by a shoe. By our testing, this would have been less than one minute with the standardized indoor conditions (Figure 27).

It also helped to go back to the scene and reproduce what was seen in the photographs and have specific times to support our findings. In my testimony, I told the jury that I would give the defendant every benefit of doubt and double the time between blood deposition and disturbance. I would give it two minutes.

The defendant was found guilty again and sentenced to prison. The defense did not offer any other experts for testimony.

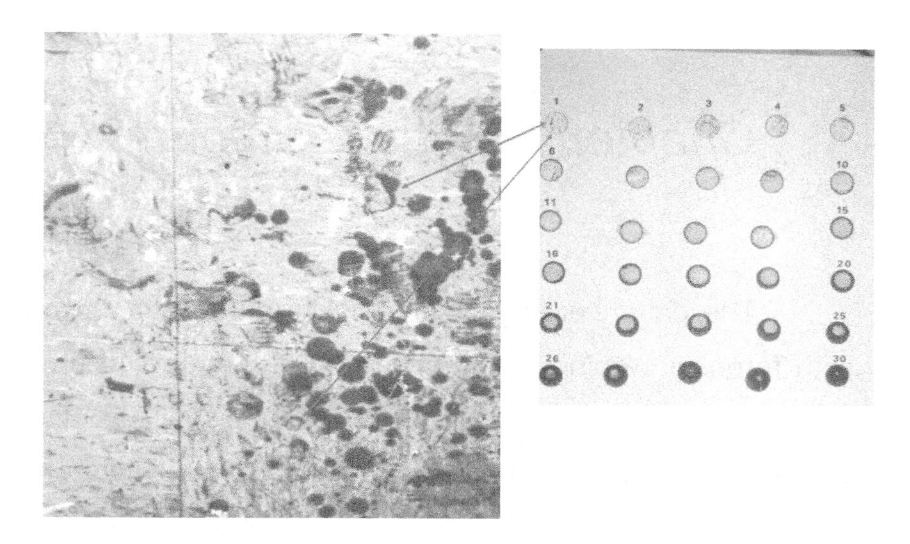

Figure 27

A few years later, I presented this case at the California District Attorneys Association Tenth Annual National Homicide Symposium in San Diego, California. There were over eight hundred attorneys and investigators in attendance at the conference, enough to give me a bit of the jitters when I stepped to the podium.

Since that case, we have had numerous other cases where time was a factor when looking at blood. In a criminal investigation, you ask, "Who, what, where, when and by whom?" With this new science, in some cases, we could offer the *when*.

Chapter 15 The Promotion

The days were flying by quickly in 1989. I was busy with forensic serology, crime scenes, and wrapping up the Francke case. Then one day I got a call from the superintendent of state police. That was like a soldier getting a call from a five-star general. The words "Yes, sir" were used multiple times during the conversation as he was telling me I needed to go to a certain town on the coast and reprocess a vehicle. "Yes, sir, did you say 'reprocess'?"

The vehicle had been processed a few days prior by a lieutenant from another crime lab. Holy cow! I was a working stiff, and he wanted me to reprocess a vehicle already processed by someone higher up the command? After the final, "Yes, sir," I ran into the office of my boss who was also a lieutenant, quickly explained the telephone call, and said, "I am not going without you." These directives were supposed to follow the chain of command. Getting a call from the superintendent was new. There was no time to ask, "Why me?" but I must admit that on reflection, it did make me smile. The other person had processed the vehicle under the watchful eye of a local detective.

The detective did not like what he observed in the processing and the conclusions reached by the crime lab lieutenant.

My boss and I went to the small town and reprocessed the vehicle together. There was reported to be a bloodspatter pattern on the front console. This was a problem, in that there was no one reported to have bled inside this vehicle. I found the pattern but testing for blood was negative. The other person did not conduct the routine test before concluding there was a bloodspatter pattern present. There was a pattern from a liquid, and I told the detective that if I were to guess, it looked like a spilled soft drink. Sure enough, through a follow-up discussion with the client, it was reported that the driver had spilled a soft drink only a few days prior. I did not test for sugar, which would have been helpful, but the outcome was going to be the same.

Apparently, the local agency head had contacted the state police superintendent's office about their concerns with the initial search. How I got the call remains a mystery. In light of my findings and other issues, the other person was demoted and transferred. This left an opening for lab director for the lab on the coast and certainly piqued my interest.

I had been suffering from local Willamette Valley pollen and needed to get out of there if I was going to survive. I applied for the position and received the promotion and transfer. With my involvement in the prior case, I felt a little uneasy about it, but apparently upper management did not.

The new position was perfect. It required supervising the lab staff but was also a working position in which I still did casework and responded to crime scenes. My health immediately improved with the fresh salt air, not to mention that I like to fish, and offshore opportunity was only five miles away. The new staff was great, and we functioned well as a group. The crime lab division had divided up the state according to geographic location, and I set about meeting the agencies and offering training if anyone was interested. Curry County was included and was remote, with the only access to the outside world by way of Highway 101. This was the only county in Oregon where a person apprehended by law enforcement with long hair and beard got a haircut before being released from jail.

With the transfer, I received a promotion to lieutenant. I had been in the new lab only a few weeks when I read in the newspaper that the body of a young girl was found along the highway and death was ruled to be natural causes. I had not heard anything about it. My subsequent research indicated that the girl was homeless and had been seen alone locally for several days with a backpack and two dogs. A call to the Curry County agency revealed that she was found inside a sleeping bag, with no dogs and no shoes.

No autopsy was performed, and no sexual assault kit was obtained. I could not believe what I was hearing, so I called the DA to talk. He was not concerned and stated that they did not like to give the county a bad image. I was speechless! How do you respond to that kind of thinking? I offered training for their law enforcement, and he accepted the idea.

I immediately set up a series of classes once every few weeks and made some new friends in the area. The classes usually started at nine in the morning. One morning I met a local deputy from Curry County at six. We went up the Rogue River, caught two steelhead trout, and got to the class on time. Life was good. It was not the sort of thing to be advertised at the time. Having a fishing pole, tackle box, and two large fish in the trunk of a state police vehicle would have been frowned upon if I had been discovered. But I wasn't.

After I had given several classes, I got a call from a deputy, who said that they had a suicide and wondered if I wanted to view the scene. I agreed and drove again to Curry County to the house. When I got there, there were plain clothed people all over the scene. I inquired who the people were and was told they were the local volunteer fire department and were searching the scene for any prescription drugs, without wearing any gloves. I tried to hide my anger. Fingerprinting was out. I examined the woman's body, and she had a single gunshot wound to the head.

I asked, "Where was the gun?" and got a blank look from the people in the room and the deputy.

"How do you have a suicide without a gun?" I asked.

This obviously was no suicide, but every drawer had been opened and the house completely searched by the locals.

I requested everyone leave the residence, told the deputy to get the body transported for autopsy, and carefully explained how this had not been handled properly. A few days later, I called the agency to find out when the autopsy was scheduled only to discover that they decided to go with the suicide. I suspect that after I left, the weapon mysteriously appeared from someone's pocket. It was clear that I had much more work to do, and I started traveling to the county regularly to continue the classes and hang out with local law enforcement. I had informed all the agencies that I was available 24-7 to talk. Call me. The next time there was a suspicious death, I got the call, and the scene was secured. We were finally making progress in this community.

Within the lab, the staff was excellent and all female, not that it mattered. Our secretary had recently given birth to a son and was running out of maternity leave.

She needed more time, and I needed her at work to answer the phones so that the rest of us could get something done. When we would talk about it, I could tell she was struggling. I asked her if she had an extra crib, which she did. I then requested that she set it up in the lab office area and bring the child to work with her. I solved my phone issue, and she was elated to have her child close where she could breastfeed him when he was hungry. I did not care what time she took for attending to the child as long as the phones got answered so that the rest of us could get something done. Everyone was happy with the circumstance until I got a call from headquarters about the arrangement. A captain was coming to the lab to interview me and the staff. It did not look good. This was outside the comfort zone of headquarters, despite no procedural infractions. My chain of command told me to seek alternatives.

When he arrived, he also had a woman from human resources with him, which turned out to be quite fortunate. I could tell from the line of questioning that he was not happy with the arrangement even though the work was getting done and all staff were supportive. Toward the end of the interview, the tone changed. The HR person, who was taking notes, found her voice. She looked at him and said, "Why not?" Every mother in the department would support this idea. What was wrong with him?

As she found her voice, he lost his, thanked me for being creative in the workplace, and left. I was never sure if he was just polite or was living in fear of the three-hour drive back to Salem with this woman. All of us could tell there was a sudden change in attitude, and high fives were in order.

As a group, we dodged a bullet, and it made us all closer. The lab design was such that there were no biohazards to the child in her work area, and eventually she would secure babysitting while she was at work.

Months later I got feedback from my chain of command that no man in headquarters was willing to take on this issue, and it died quietly. One cannot underestimate the power of women in the workplace, even in a male-dominated profession. I don't know why I would be surprised; I had three daughters and a wife who already ruled my life.

Chapter 16 Photography and Domestic Violence

Figure 28

In my early married life, I did a lot of wildlife photography and wedding photography. Weddings provided some extra income. It is no great statement that photography is an art form. One has to frame the shot, adjust the light, have the correct film, and adjust the depth of field and the shutter speed. Each item has an effect on the appearance of the final print. If you do your own developing, you can also control contrast and exposure independent of the camera settings.

There are thousands of professional photographers around, but few understand the physics behind the photography, especially if you get into Infrared (IR) and ultraviolet (UV) photography. Coming into the crime lab, I had an eye for the art form, but when I started to mess around with the old Graflex as seen in figure 28, I began to spend more time on the physics. It was important to know why you got a certain image. Here was the simplest of cameras, yet it had tremendous potential. How ironic. Every feature on the camera was controllable, so if you understood the effect on film, you could get tremendous detail with a negative that was four inches by five inches in size. The negative for a thirty-five-millimeter camera was about one inch by one inch. Size of a negative matters in getting a detailed image.

But back to these cameras, in order to focus, you had to open the back of the camera and look through the front lens to see an image the same as the film would see it. Once focused on an image, you took out a light meter, set it for the ASA or the ability of the film to gather light, and determined the shutter speed and the f-stop. The f-stop was set depending on what you wanted in focus. There are books written on how to get all this right, so I am not covering it in any greater detail.

Once the camera was set up, a film back was attached to the back of the camera, a metal plate was removed that was between the film and the lens, and the shutter was cocked manually. Then the shutter was released; the lens opened for a predetermined time, and light struck the film surface.

The chemicals on the surface of the film are sensitive to light, so this exposure caused a chemical reaction depending on how much light struck the film surface.

Unlike modern cameras, most of these cameras were old enough that they did not have a wavelength-blocking coating on the lens. The lens was also quite simple, only one piece of glass, which allowed maximum light to enter. In the modern lenses, there are multiple glass lenses within the barrel. The more glass you place between the subject and the film, the less light will enter the camera.

The film was also an area of interest. Kodak print films were called *panchromatic*. This meant that the film was sensitive to light down in the high-energy, short wavelength, UV, and at the other end of the light spectrum sensitive into the far infrared region. So the film was capable of capturing light beyond what our eyes could see. Why bother? Because physics tells us, for example, that the shorter the wavelength, the higher the resolution. You could either optimize for certain wavelengths with filters or eliminate other unwanted wavelengths, depending on what you wanted to accomplish.

In photographing human skin, it is sensitive to certain wavelengths of light. Short wavelengths reflect off the surface and show scars, needle marks, and bite marks very well. IR, on the other hand, penetrates deeper into the skin and document bruises quite well. IR is just beyond the red color of the visible region, so you get a reddish image with color film.

But there is no color in the UV or IR region, so one should use B&W (black and white) film. The physics is good to this point, but what about all the stuff between the film and object? You must be able to focus with available light, and you need an IR source.

I found a red lightbulb to be a good IR source. To get rid of all of the other wavelengths of light, I tried several different filter combinations, but each one reduced the amount of light I needed for the wavelengths that I wanted.

Then it occurred to me: One simply needed to turn off the lights in the room to get rid of unwanted wavelengths of light! Modern cameras had a coating on the lenses that prevented wavelengths outside visible region from entering so that photographs would be true to color, no excessive blues or reds. According to the Kodak technicians that I talked with, this might also include coatings on internal lenses, too. Each time photons entered another piece of glass, some would be lost. Hence the Graflex with one piece of glass in the lens had a distinct advantage for UV and IR photography, when the amount of light generated would be limited. If photons were lost, it would be in the space between the lens and the film plane.

How did I know what lenses worked and what did not? I altered the chamber of a spectrophotometer so that I could place a camera lens inside and get a spectrum of the light passing through. Now I knew exactly what the film was seeing when the photons hit the film plane.

The OSP was well aware of what I was doing and offered encouragement. An environmental case from the Columbia River area presented itself with a number of fiftygallon drums of a toxin that were tossed into the river to get rid of them. OSP had recovered them and now needed to know where they came from. I got a call from division wondering if my photo methods might help identify the company of origin, so I talked to the OSP investigators working with DEQ about the case. They said that the labels had been altered and they could not see the numbers that identified the company that originally had the barrels. Apparently, the substance was highly toxic to fish in the river and capable of doing serious damage if the contents had leaked into the river. There was no one else capable of recovering the information, and they wondered if I could give it a try utilizing specialized photography.

I requested that they remove the labels by cutting out the metal of the barrel without applying any heat to the labels.

I received them a few days later and could not see the numbers, either. They were just too weak to identify visually. I set up my Graflex and started working with several different filters in combination with UV, visible, and IR light sources. I was able to locate and identify every letter and number on the labels using reflected UV. The originating company was identified and was later convicted of environmental crimes based upon my success with that old camera.

As I recall, the company had to pay fines in excess of $1 million. Two photographs are shown below that depict the original visible light image (Figure 29) and the label through UV photography (Figure 30). There were accolades all around, and I felt on top of the world for a few days. All my tinkering with this old piece of equipment finally paid off. I suspect the camera was as old as I was at the time.

Figure 29

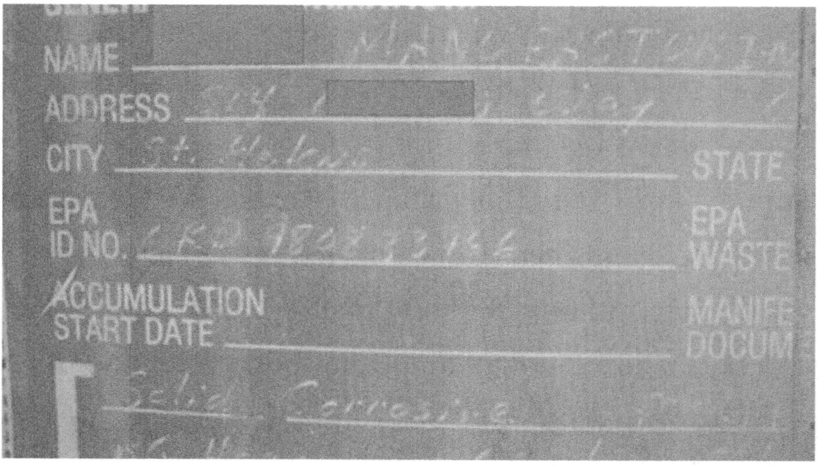

Figure 30

A few years later in the early 1990s, I got a call from the DA in a neighboring county about a child abuse issue. A young girl was reportedly beaten by her father. But time had passed before the reporting, and bruises were no longer visible. I told the DA that I would make no promises, but I needed to see the girl. She was transported to the lab the next day, and I interviewed her and examined her ankle for any bruises. This was the last place on her body that was still painful. Take all the complexities of working with a Graflex and now add in a moving target, and this really got interesting.

I could place the camera on a tripod and focus on her leg, but she would move by the time I got everything set up. The photo was blurred. In this low-light-intensity environment, my depth of field was of a fraction of an inch, and I had to focus without any film, add the back, turn out the lights, cock the shutter, turn on the UV, and release the shutter. Time for a new idea. I took a wooden dowel and taped it to the bottom of the camera with the end of the dowel being about two feet from the lens. The focal distance was adjusted to match the end of the stick. Then I turned out the lights, turned on the UV light, poked her with the stick, and hit the shutter. It worked, and I got a rough image of a possible bruise on her ankle in an area of light not visible to the naked eye. Poking young girls in the dark with a stick was a new touch, so to speak, for our lab system. The Polaroid picture showed a lot of surface detail, but a light image of an apparent bruise came through.

I followed up and got the best result with an IR source that I could expect and called the DA that I had an image. I was unprepared to testify that the object in the image was a bruise. He was disappointed, and the conversation ended there. I was no use to him. The next day the judge in the case called me. He wanted to hear for himself what I had seen. I explained it to him and said the image was in the area she had described as painful. There just had not been enough research done in this field, and I needed some time before I was comfortable in testifying about it. He did not care about that; he was satisfied that there was something visible that was "potentially" a bruise. There was other evidence in the case, and this was supportive. The dad was on probation, and the judge sent him right back to prison.

Word spread around the agencies that I had a new way of documenting bruises in domestic violence cases, and the phone started ringing. I did not realize that the use of photography in abuse cases was so limited and without any uniform procedural documentation. This was an area of law enforcement that is a regular part of daily activities, and no one had developed a standardized procedure?

I needed some research time; anyone in the office or in my family who had a bruise got photographed. The problems included using the right filters, optimizing the camera, and knowing what film to use and how to do a good interview. It quickly became apparent that the interview had to be a hands on discussion. Many of the reports that I read relating to DV victims were woefully inadequate in documenting the injuries to the victims or the circumstances of the assault. During an assault, there were often readily visible bruising, but the victim often had injuries that he or she was not aware of, due to the heightened emotional state of the mind during the assault.

I developed a procedure where I would start at the top of the head and work all the way to the feet. Putting my hands around their head, I would simply ask them to tell me if anywhere I touched hurt. From there, I'd move on to the neck, both arms, and their feet. A person of the same gender as the victim would first look over the torso to determine what if any bruises were visible.

If they were present, we would cover the sensitive areas of their body, only exposing the bruise or, on occasion, a bite mark.

With good documentation, the success in prosecution was going up. The number of victims requesting photography was rising rapidly, and we were doing several people a week.

I held a meeting with several detectives and DAs involved in domestic violence prosecution just to get a handle on what was going on currently. It was a disaster. There was no consistency in working with victims; nothing was set in advance in preparation to handle a victim when the crime occurred. Not only were the injuries to the victims not being documented but no one was photographing changes to the scenes also. This might include overturned furniture or broken glass, as examples. Most officers in the area did not know how to use the cameras they had available. In one instance, the officers were ordered to use just one camera for the department and to limit the number of photos taken. The roll was left in the camera until all twenty-four photos were taken. Then the film was replaced, and the exposed film was labeled and placed in a drawer. No prints were made unless there was a specific request for them. Often the DA did not even know there were potential prints available.

The current concept of a male officer photographing a female victim was a problem. Some of the stories I heard would have been funny, had this not been so serious. In one instance that I heard of, the female victim bared her chest to show the officer a bruise on her breast. He immediately ran out the door as if he had been shot. He located a female dispatcher to take the photos for him, as he was literally afraid to continue. She had never seen or used the camera before and had no skills beyond point and shoot. Consequently, the photos did not turn out, and the evidence was lost.

Besides the photography, I had to attend a crash course on domestic-violence and interview procedures with children. In my line of work, usually we only saw the worst cases in which the abused did not survive. In the other cases where the victim survived, the problem is best described in the cycle-of-violence scenario: An argument develops, the altercation turns physical, and someone gets hurt. The crime is reported, and the officers take a report. If the issue was serious enough, the abuser went to jail.

The problem was the other side of the circle. The abused and the abuser reconciled their differences, and everyone was happy again, at least until the violence was repeated. In time, the abused would call the DA and request that the charges be dropped. Without the abused as a witness, there was no case. One cannot blame the victims; they may be entirely dependent upon the abuser for housing, food, and any other needs. Survival on their own may not be an option. Periodically the cycle would repeat itself again and again.

As a result, the conviction rate for these offenses was less than 10 percent. The low success rate caused the officers in the field to make these cases a low priority, and not much effort went into them.

There was a lot of work to do, and it started with writing a procedure for DV investigation and photography to be distributed to the local agencies. I went on a training blitz and taught the officers how to use their cameras, an important first step. When they had a case, they would complete a form and send it to the DA's victim-assistance office so that the DA knew there was someone out there that needed attention. The next day victim's assistance would call me and schedule an appointment to bring in the victim. One of my staff or I would do the interview and the photography. While I was getting familiar with the photography problems and writing procedures, I had to develop a technique that the officers could do with their own cameras. The procedure here had to be very simple, or it would not be received well by the agencies.

The photographs and reports of the cases I was working in the lab were forwarded on to the DA's office, and suddenly things were changing. A deputy DA was assigned to work with me on these cases as the numbers were jumping rapidly. No one had addressed domestic violence cases so thoroughly in the past. The prosecution success jumped to 72 percent! The photographs now had multiple uses.

The judges were using the photos for sentencing. Jailers were using them to determine if they were going to release the accused from jail and, of course, as evidence in criminal prosecution. Often the DA's office could get the accused into counseling by telling them that if they were caught again, they had the photos from the first event to add onto the new charges. Now the officers in the field were interested, because something good was coming out of these cases.

The recidivism for these crimes was normally very high because of ineffective prosecution. But in over five hundred cases that I personally handled over a span of five years, I only had one person come back a second time. It turned out the first time was for domestic assault, but the second time was a sexual assault unrelated to the accused in the first case. A nonexistent recidivism rate is unheard-of for this type of crime. For the first time, for abused women who came to us, lives were beginning to change. There now was a motivated deputy DA and a new staff of victim-assistance people who were excited about the success rate.

There was daylight out there, and a new life was possible free from abuse. The abusers were being held accountable due to good physical evidence and a promise of harsher penalties if they did not stop the abuse.

In my twenty-plus years with the Oregon State Police, I had opened doors in a number of areas, had some great success in casework, and was now the director of a crime lab. But drugs and domestic violence were the central issues in the origin of crime. For the first time, I had developed something that went to the heart of crime and had a direct impact on the living victims. It did not seem that difficult, but it turned out to be one of the most important contributions to forensic science and my local community in my career. The OSP bought into the program, and we had federal grants and specially trained technicians to document the injuries of DV victims just as I had been doing. Requests for lectures all over the United States were coming in, and a national television network did a special program on our procedure. Occasionally a woman who was a prior victim would approach me while off duty out and about in the community and thank me for how I had helped change her life. For me, there was no greater reward.

The process I developed was applicable to any police agency and followed the basic steps as follows:

1. Document through photography any alterations to the scene associated with the assault.
2. Have a preapproved process in place to handle DV victims, usually someone in the DA's office. Schedule an appointment for photography of the injuries within a few days.
3. Begin the interview by placing your hands on the person's head and pressing fingers in search of tender areas. When located, ask why the area is tender and how the injury may have occurred, if the

person remembers. It is OK if the person does not remember.

4. Move down the body, including extremities, to the toes and search for injuries and responses to those injuries.
5. Photograph each injury with the camera set at the minimum focusing distance.
6. Circle the bruise and re-photograph it with a scale.
7. If possible, re-photograph the injuries using enhanced photo techniques, such as IR and UV or an alternate light source.
8. Do not use a flash or available fluorescent room lighting; use an ordinary lightbulb for the only light source.

Step eight was interesting, in that a camera flash was too bright for close-ups and overhead fluorescent lights common to police stations and hospitals emit too much light in the blue region.

An ordinary lightbulb emits more light in the red region, and better results were obtained.

We were making a difference, and the media was busy reporting our results. The department was also using my work in the budgeting process to get more funds. The young deputy DA who was working with me on these cases used her background in this program to apply and get herself appointed as the US district attorney for the state of Oregon. For me, good things came from diligence despite having to keep up with my other lab responsibilities. Shown below in Figure 31 are a series of photos that demonstrate the improvements in visualizing bruises. A technique that I developed.

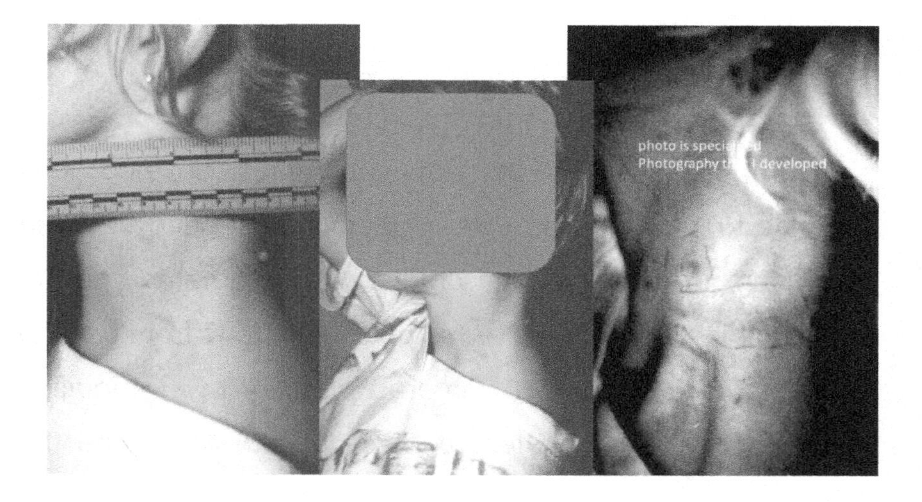

Figure 31

A few years later, we were seeing increased numbers of sexual assault cases in which the victim was bound but no visible proof of bonding was available by the time law enforcement became involved.

Any marks had simply disappeared by then. I wondered if there was a way to document the marks through UV photography and the length of time the marks might remain.

If I could document the marks out a couple hours, there was a chance to prove bondage.

I had demonstrated before that UV photography on skin was a surface technique. The high-energy photons reflected from the surface without much penetration. If there was enough time, this method of photography should be able to document indentations left on the skin from the bondage item.

In regular photography, UV light will turn your photographs blue, so modern companies develop lens coatings to prevent that from happening. I needed to make this technique possible with ordinary SLR cameras and not anything highly specialized. So I placed the lens from several different brands of cameras into my modified spectrophotometer and tested each lens to determine the extent of UV light passage. I found that the coating on Minoltas and Canons did allow light down to UV. This was enough to get a sharp UV image on B&W film if the light source was close to the lens.

Getting a quality photograph meant control over several aspects of the photographic process. One had to know what wavelengths of light to use, what wavelengths come from the light source, what wavelengths will pass through the lens, and what wavelengths was the film capable of recording. For example, as mentioned before, a modern camera lens in an SLR has a multitude of glass lenses within the housing; each one might have a UV resistant coating. Each time you pass light through another piece of glass, you lose some intensity.

If you lose intensity, that requires a longer exposure time. Longer exposure times lead to blurred images with only slight movement when the lens is open. Therefore the simpler lens with the least amount of glass was preferable and that is what made the old Graflex such a useful camera.

But where do you get subjects to photograph? You need controlled studies before you go for an actual case. The next morning after testing the cameras I brought some three-strand rope to work and talked to my secretary about being a volunteer while holding the rope. This was not the first time I had asked for volunteers, and she was hesitant without more information. A little explanation and persuasion, and we were on our way. As she sat at the front desk in front of other office staff, I tied her wrist with the rope tight enough to dent the skin but not so much as to be painful. I left the rope in place for twenty minutes, which drew a lot of laughs from the rest of the staff. My secretary and I were now into bondage.

After twenty minutes, I had the secretary come into a room where I could turn out the lights. Again more laughs. Turning out the lights removed all the other wavelengths of light except what my UV source was providing. I placed the

UV light directly on the back of her outstretched fingers to get source as low as possible and still maximize the intensity. A light source will lose intensity exponentially as the distance increases.

The camera was set perpendicular to the wrist at the minimum focusing distance of about eighteen inches, and I started shooting. Keep in mind that I did not want to use a macro lens, as that would be more glass. It took several tries, and I took the photos with both visible light (lights on in the room) and with only the UV light source (lights out). We took photos every fifteen minutes for about three hours until I could no longer see a pattern.

To the naked eye, the pattern on the skin was gone in thirty minutes or less. To put this problem in perspective, one had to consider the chain of events in a crime. The bondage occurs, the bonds are removed, and the person has contact with law enforcement at some point and is often transported to a hospital for exam. Three hours is not a lot of time, but thirty minutes was useless. Three hours was within the window of opportunity for documentation of the bondage in some cases. I mentioned the technique to some detectives and asked them to watch for an opportunity to document bondage in a case, but one never presented itself before I retired. Years later I was having a conversation with Dr. Michael Baden, the famed television personality, and showed him the photos that I had taken. He was very impressed; he did not know of anyone who had perfected this technique and saw the immediate application. In many sexual assault cases, the issue of consent of the victim can be a difficult hurtle to overcome. The demonstration of rope marks on the wrists can go a long way to support the victim's statements that she did not give consent.

I often thought that I should publish my findings in the *Journal of Forensic Sciences* but just never got around to it. An opportunity missed for me, but perhaps someone else will pick up the torch and work this up. The photos below show images of the skin taken with a single-lens reflex camera suitable for doing both visible light and reflected UV light photography. As mentioned before, the UV is accomplished by turning out the room lights, turning on a UV lamp, holding it at about the palm of the hand to lower the angle of light, and taking a photo. The visible light photos were not focused well in figure 32, but there were no marks visible to the eye after about thirty minutes. Three hours may be long enough for a rape victim to get to someone who can document the marks supporting the statement that the act was against his or her will.

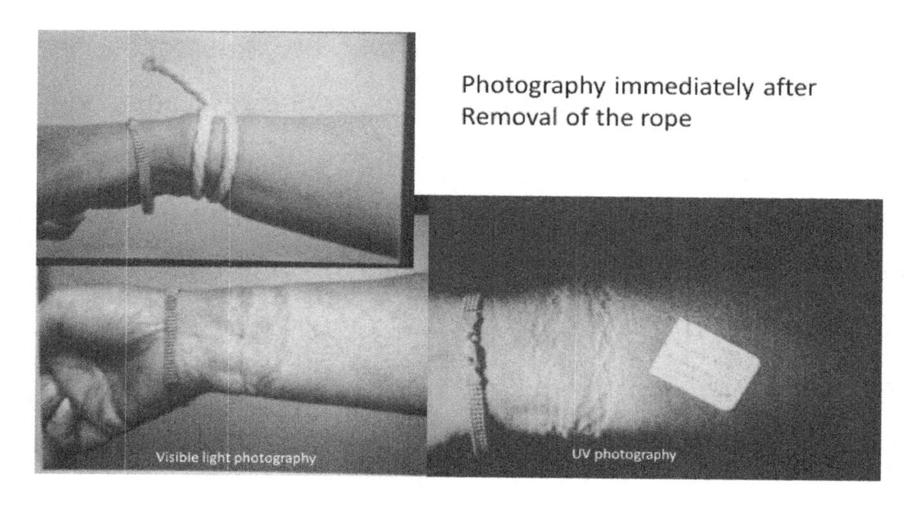

Figure 32

Photography of roped area 2 hours later

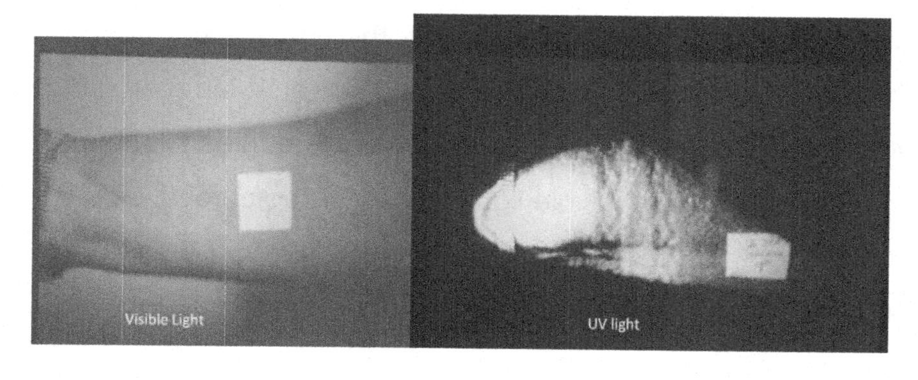

Figure 33

A few years into the domestic abuse project, I got a call from an OSP station commander in another region of the state. One of his troopers was involved in an altercation with a rather athletic woman who had resisted arrest.

She got lose from his grip, ran for her car, and grabbed a black object. It was dark, at a roadside in a rural county. The trooper saw the object, reached for his pistol, and shot her. She was wounded and later transported to a large hospital in Portland. He was injured from the altercation as well.

The station commander wanted me to drive to the small town in eastern Oregon to interview him and photograph his injuries. What he did not know was that I had broken my leg a week prior and was on crutches at home. I could not drive and, because of the pain, could not sit in a vehicle for several hours, either. I suggested he needed to find someone else, as I had trained several people in the state to do DV photography. An hour later the phone rang again, and he said he had a plane coming to pick me up. This was winter, and there was snow on the ground where I was going. I initially objected; as an excuse, I told him my crutches did not have any traction tread. He laughed, and after discussion, I finally agreed. Jennifer, one of my techs would carry the equipment.

We flew to a small town in southeastern Oregon. We met at the patrol office, and I did an initial interview of the trooper. The trooper was sore, but there were no bruises showing anywhere. I suggested that we have an ER physician look at him. I believed the ER doctor would make a good witness to his injuries despite the lack of visible bruises.

In the place of bruise photography, I asked the doctor to use a black ink marker and outline the muscle groups that were affected. I would photograph what he outlined. This was a new approach. I had not done this before.

All in all, it took several hours to get the interview completed, and my leg was hurting. I requested someone take us back to the motel so that I could put my leg up and take some pain medication. Unfortunately, this was not going to happen. There was a call from headquarters, and I was to be flown directly to Portland to interview the victim. A broken leg was of no consequence.

Despite my present condition, I found the flight over the Cascades beautiful. A trooper in a patrol car was waiting for us at the airport and transported us to the hospital. It was late night by now, and my leg was killing me, causing me to grit my teeth because of the pain.

We got to the woman's room, and she was sitting in a chair, heavily bandaged, with her lawyer standing right behind her. I said hello and asked how she was doing. There was no place to sit in the room, and no one was offering me a chair from another room. I said, "If you don't mind, I am hurting so much that I will sit on the floor," and handed my crutches to Jennifer. The woman was uncomfortable, but she could see that I was in pain, too. Every time I asked a question, the lawyer would interject that she was not answering any questions. I gave up.

I said, "Do you mind if I just sit here a minute? I have been all over this state today and am tired, and my leg hurts."

My leg, which was in a cast, was so sore that I could not stand without assistance. We talked about my injury and the way that I'd had to crawl two hundred feet off the side of a cliff to get help. It was not intentional, but I could sense she was softening and showing some compassion for my condition. I told her I was sorry that I had visited her in my current predicament but was ordered to come and ask her a few questions. I asked one last time if she could just tell me her version of what happened. The lawyer interjected again with an aggressive voice and again said no.

To my surprise, she turned her head to the lawyer and told her to shut up. Then she began to talk and tell me about the incident. I took notes and thanked her again for her willingness to see me. From there, it was back to the airport, a flight to my home city, and back to bed. I had been on the go for over twenty hours and was exhausted. I took the rest of the week off, there was no telling what was in store if I went back early. Months later the courts ruled the shooting was justified; the photos I had taken along with the interview were used as support for the case.

After that, the case went civil, and the Department of Justice, the agency that handles civil cases for state agencies, was sufficiently confident of their case that they did not use me, my report, or my photos. The state lost, and the woman was awarded several million dollars by the jury.

Chapter 20 The Michael Francke case

Late one evening in January 1989, the superintendent of state police called my boss and directed that Mike and I travel to Salem for a homicide investigation. The head of the Department of Corrections was dead on the portico of the corrections building. His name was Michael Francke.

When we arrived, it was like the who's who of law enforcement. I think the entire headquarters staff was present and the DA's office staff as well. Michael Francke was a friend of the governor and a highly respected public figure in state government. He had been hired to clean up the corrections division, as there were rumors of corruption from within the department. Mike and I got out of our van and walked up to the superintendent, who was talking to the governor. We identified ourselves and kind of stood around, not sure where to start or knowing what had been done thus far. We needed a briefing. There were blue uniforms everywhere and folks rushing about with assignments to complete, but we had no idea who was in charge of the scene. Then I asked a kind of necessary but foolish question: "Has anyone searched the corrections building for the suspect? Are we sure he is not on site?"

There was a long pause. Then the superintendent starting shouting orders, and more people were quickly moving toward the building. Keep in mind that this was a huge, multistory state building with 180 rooms. That did not include the closets and housekeeping areas where someone might hide. We were assigned areas, and all of us began the search with weapons in hand. My group entered the building through a side door and began the room-by-room search. It was announce and advance as you went. Somehow, I got separated from the group and was on my own and about half-spooked. In one room, I quickly opened the closet door, and there was a mannequin. I nearly shot it, as my heart rate hit Mach two—a lucky day for the mannequin. Hours later the search was complete, and we reconvened in the parking lot in the daylight. No suspect was found.

No one was really sure what had happened. Francke's car was unlocked, and there was a blood trail from the parking lot and up the front steps that ended at his body. A windowpane was broken near the door handle with glass on the inside, indicating that the break-in was from outside to inside.

Francke was on his back in front of the door with a knife wound in his chest. As soon as I saw him, I asked, "Who moved him?" I knew the firemen and the EMTs had been on the portico to establish that Francke was dead. And as I suspected, they had rolled him over to check for a heartbeat. Initially no one wanted to fess up or had notes that supported that action.

We secured the area and began processing the immediate scene around the body. Several photos were taken, and we removed the outer clothing at the scene. There was a long dark head hair on the scarf around his neck and several shoe impressions on the portico.

Messing with the body at the scene is always a point of contention with medical examiners (ME). Historically we were not allowed to remove any clothing until the ME got a chance to see the body, which was usually on the autopsy table. But I had had a couple cases where I could see important evidence on the body, placed the corpse in a body bag, and sent it off for autopsy, and when we opened the bag, the evidence was gone. If it was not for the fact that we were required to process the scene and go to the autopsy, no one would have known. The final straw for me came when I was at an autopsy standing next to the body with an evidence bag waiting for the ME to remove a shirt. As soon as he took it off, he gave it a good shake to straighten out the wrinkles! I nearly went nuts! There went all the physical evidence!

So I started stripping the bodies at the scene. The first time I got scolded by the ME, I told him that the clothing was in the lab and he could stop by and see all of it. He came by the lab, and I had it all laid out on butcher paper The ME made a reach for one of the items, and I stopped him. I said, "If you touch it, I will need hair and fiber standards from you."

He backed off and did not give me any more guff about taking the clothing. We had reached an understanding on the value of trace evidence.

At the scene we noticed that Franke's hands were cut, possibly from breaking the glass in the door. After a good topical examination, we assisted in packaging the body, which was sent to Portland for autopsy. There we learned that Michael Francke had a single perforating knife wound to the chest that penetrated his heart. Our suspicion was that Francke was stabbed in the parking lot and traveled about one hundred feet to the door of the portico before he collapsed. If you watch enough television, you would think that an injury like this would be instantly fatal, but it is not. Death requires oxygen deprivation to the brain, and that could take a few minutes. In this case, there was enough time to reach the door. Unfortunately, the doorknob was broken and would spin without opening the door. It appeared that he tried to open the door from the outside by breaking the window before he collapsed.

After finishing up on the portico, Mike and I organized a search of the entire grounds. One thing we had available was lots of help. There were still blue uniforms everywhere.

We set up a line search and instructed the officers to pick up anything that looked out of place. But what we needed most was a knife. We continued with our work and noted that between the parking lot and the portico was soft soil; if you stepped off the walkway, you could leave an impression. And there, sure enough, we could see shoe impressions.

These were measured and photographed. A mix was prepared, and the more significant shoe impressions were cast. With fifty or more people at the scene potentially creating shoe impressions and consideration for a suspect as well, the impressions were going to be a challenge to identify to a specific shoe.

We collected several shoe impressions on the grounds near the building and many more shoe impressions on the portico. Considering the scene, I had a hunch. My first question was, who was here before the scene was secured? We needed shoe exemplars from everyone who had been on the portico or directly in front of it. When I told that to the superintendent, there was a moment of silence. I looked at him and already knew what he was going to say. He admitted he had been up there. In fact most of his personal staff had been up there on the portico too. OK, so a few days later, Mike and I got shoe exemplars from every one of the trespassers. They knew better but thought they would not be discovered. One needs to eliminate all the shoe impressions possible; those that cannot be eliminated could potentially be the suspect.

Mike and I did not finish up the scene until late afternoon. By then the superintendent had called every OSP detective in the department to Salem, at least fifty of them. A conspiracy theory was floating around that there was drug activity at the penitentiary and that Francke was about to expose it, so the detectives had plenty to do.

During the day, Mike came down with the flu and spent the afternoon in the back seat of our vehicle periodically opening the door and vomiting. He was not doing well, but we still had a job to do. He was out of sight in the back seat of our vehicle, and I kept going.

After the initial scene, we went to the Francke personal residence that night and searched it looking for any investigative leads. The search of the house was uneventful, and we returned to the lab late that night with evidence from the scene. Mike went home and took a couple of days off. I was bone-tired; we were running over thirty hours without sleep.

After Mike recovered, we returned to the corrections building after dark to luminol the area. We arrived about ten at night and discovered that there was a large halogen light illuminating the area we wanted to search. We started making phone calls and could not locate anyone who knew how to turn the light off. One officer from Salem PD who was standing next to me simply pulled out his weapon and aimed at the light, ready to shoot. I yelled to hold it a moment before we got that drastic.

With the recent homicide, I couldn't imagine the attention we would attract if he shot out the light. His only comment was, "Seemed like a good idea at the time." We finally contacted the local power company, and they shut off the light. At that point we were able to spray the parking lot and the sidewalk leading to the building. The only blood visualized by this method was already identified and probably left by Francke.

On the portico, a few of the bloodstains had a partial shoe impression visible as seen in Figure 34. This was important because there is a time element here. Based on our knowledge of blood drying times, the impression could not have been too long afterward—say within a half hour. The shoe impression had to have been made after the blood hit the floor, and the person who made it was not Michael Francke. His shoes had a different pattern. At this point the potential for the shoe impression to belong to our suspect was high. The question was, could we determine, using the environmental conditions that night, when someone stepped on the blood?

Figure 34

We knew that the night watchman had discovered the body almost four hours after we believed the incident occurred. There was no information of anyone else being on the portico any earlier.

To make it more difficult, the pattern of the shoe was similar to the night watchman's shoes. Was he lying about when he found the body? Suddenly he was a prime suspect. So how does the blood remain soft enough to receive an impression for that long, if he was telling the truth? We had no answer, and if it was not from the night watchman, could it be from the assailant with a similar shoe? With fifty detectives looking for a solution, you did not want to be that night watchman.

As I have mentioned before, blood drying time is based on the loss of water in a bloodstain according to Henry's law of physics. The only variable should be temperature. But this was January in western Oregon; it was cold and humid. We had conducted experiments in the lab, and the bloodstains of similar size dried within a half hour indoors. After a droplet dried, no shoe impression was possible given the circumstances. Mike had a flash of insight, and we checked the NOAA weather data for the evening of the incident and waited for a similar opportunity. We had never done any testing for drying time in these extreme outdoor circumstances. We ended up waiting until the following winter before we could test blood directly on the portico. That night had finally arrived when conditions were right.

We drove from Eugene to Salem and started placing blood droplets on the floor of the portico. At five-minute intervals, we would blot the bloodstains to see how much had dried. The process was depicted in Figure 35.

The outside temperature was in the forties, and it was foggy. That meant we were at the dew point and humidity was virtually 100 percent. In this environment, Henry's law has a problem, in that the air is saturated with water. There is no more room for more water in the air above the droplets when humidity hits 100 percent.

Figure 35

Now an interesting phenomenon occurred; the blood did not totally dry. This was an observation that we had never seen. After an hour, the blood went into a tacky stage where the blood was soft but not completely dry. We were not sure how long this condition would last. But we watched it for four hours, and it remained the same consistency. This was new information and new science. No one had reported this phenomenon before. But it did explain who may have left the shoe impression, and the night watchman remained a possibility. He had shoes that fit the class characteristics of the impression in the blood, and the blood remained soft enough to accept a shoe impression. One problem was reasonably answered.

While we were on the portico that night, we had a visitor. Michael Francke had two brothers, and one of them walked onto the portico when we were doing our experiments. We quickly struck up a conversation, and I offered to show him the new crime scene van. He and I left the portico and went to the van before he could realize what we were doing. I held his attention for almost an hour while Mike finished up the testing.

The Francke brothers were friendly to the press, and we did not want him to know what we were doing.

The *Oregonian* and the *Statesman Journal*, newspapers from Portland and Salem respectively, had been covering the case from day one. They were always critical of the investigation, and when a suspect was developed, they were certain that he was the wrong guy. Working on the case daily, I knew what was happening. When I read about it in the newspapers, it was hard not to call them up and tell them that they had no idea what was going on and the confidential informant information provided was false. Perhaps it was an early version of "fake news." It now reminds me of the 2016 election. What is truth?

The blood trail from the parking lot did not really provide much information other than direction of travel. The amount of blood increased as Francke reached the door of the portico. Mike and I would often have to explain the process that supported what we saw at the scene. When a person is stabbed, bleeding is immediate, but that does not mean the person will leave a blood trail. First, the clothing becomes saturated, and blood runs down to the lower edges of the fabric. Once the blood accumulates on the bottom of his pant legs and the bottom of his coat, it will dislodge, especially with movement. This takes time, and time translates to distance from the vehicle.

We could not tell if he had walked or run, only that all the bloodstains were in the direction of the portico. We did not see any indication that he had fallen until he got to the door. I am always amazed at how a simple action like blood dripping can turn into a lengthy scientific explanation.

Even with Mike's help, the months were passing, and I was getting behind again in my other casework. I appealed to my boss; he contacted headquarters, and for the second time in my career, I was relieved of all other tasks and told to just focus on this case. I know that the superintendent felt his reputation and that of OSP was on the line trying to solve this case. The media was not giving them any slack either. I was proud that the superintendent had picked me.

As the days and weeks passed, we were required to attend the frequent investigators' meetings in Salem to keep up to date. The DA's office decided they needed a diagram of the portico. It needed to be highly professional. We did not have any software to do that, but we found software available for purchase that would work. Despite the pressure, the department balked when they heard the price. Mike called the company and told them why we needed it. We got lucky, they sent us a free copy. We went back to the portico, laid down measuring tape, measured and took photographs of every brick on the portico. This was no small task and took days to complete.

In the photo below, Figure 36, shows the blood trail and is accurate to the correct brick as it was on the portico. Shoe impressions were also identified to their specific location and brick. Bruce, one of the other guys in the lab, was assigned the graphics stuff and did a yeoman's job with the new software.

Can you count the number of bricks in the graphic of Figure 36? We did. It is the same as in the photo on the left.

Figure 36

Now imagine that you have fifty detectives looking for suspects in the community and the media was providing daily editorials on the sloppy investigative work. There were sure to be some suspects developed. When this happened, a search warrant was obtained, and the home of the suspect was searched.

They were looking for clothing with bloodstains that might type back to Francke and also knives. Instead of looking at the clothing and other items on scene, they got a pickup and emptied the house of every item of clothing and trucked it to the lab. They would drive up to the back door with hundreds of clothing items, including from wives and children, and expected to drop them off and drive away.

Can you imagine someone coming into your house and taking every item of clothing that you and your family has?

I looked at the contents of the first truck and said that this was not going to do. I could be weeks looking at this stuff. I called the lead detective and told him to send me at least seven or eight detectives to assist in looking at the clothing. Knowing that the superintendent had my back, here I did have some authority. Several detectives arrived, and I set up a long table in the lab near the back door. One or two detectives took a clothing item out of the truck, another wrote a description of the item, and another took a photograph and then passed the clothing on. I sat on a stool about midway and watched the item go by, having a detective turn the item around for me so that I saw both sides. If I did not see anything, it went by me and on to another detective, who placed it in a plastic bag for return. If I did see something suspicious, I noted it, and another detective filled out a form for submission to the lab and properly packaged the item.

This process happened about a half-dozen times during the investigation. We would go through a pickup load of clothing in about four to six hours.

As on the first day, the detectives were also looking for knives that might be the possible murder weapon. Actually, everyone in the community knew we were looking for knives. One day a couple detectives were parked at a business talking to the owner, and when they returned, there was a knife lying beside their vehicle. And they submitted it! Finding blood on a knife from a single stabbing is difficult. As the knife is withdrawn, the skin will wipe it to some degree. Passing though clothing will do the same thing. I doubted there would be blood, but we had to look anyway. Over the next several months, over four hundred knives were delivered to the lab for examination. They were coming from all over the Salem area. Mike and I were getting really good at knife identification. If my memory is correct, I believe we looked at about 400 knives.

It was summer now, and I was on vacation camping with my family when the beeper rang and it was the lab. When I called, they said they had a knife that had given a positive presumptive test for blood and demanded that I return immediately. My family was none too happy about this, but that was what we did. I walked into the lab that afternoon, and a crowd was waiting for me. There were high expectations for this knife. I picked up the knife, looked at it quickly, and said, "Nope." It was not the knife that stabbed Francke.

I walked out and got a cup of coffee. Everyone was bewildered about how I could tell so quickly when no one else could. I let them think about it for a few minutes. This was easy; when I looked, there was a fish scale on the blade! It was fish blood! Fish blood gives a positive test for blood too. Later I found out that one detective, who was not present, had actually taken the knife out of a tackle box. I was not happy about giving up my vacation over this. But the case was going nowhere, and management was getting desperate.

I did have Michael Francke's clothing, and I did pick some hairs off the scarf and his other clothing. Hair comparisons have always been controversial, and in recent years, they have been reviled as total junk science. Not so. Hair comparison has value for elimination purposes and limited value for implication of a suspect. Problems happen when too much emphasis is placed on the implication. Early literature on hair comparisons attempted to apply a likelihood ratio to a matching result of standards from a person to unknown hairs from a scene. None of us in the lab bought into these likelihood ratios, and we often discussed the proper terminology for when we could not tell the unknown hairs from the standards. Here was the problem; the common term based on the FBI was *similar and consistent to a common origin*. We did not like this. It implied that the hair came from that person when in fact it could have originated from several different people.

I adopted the terminology *indistinguishable from the supplied head hair standards*.

That was the truth if you could not tell them apart. But never really satisfied with that either, in later years I used *unable to eliminate*. Another true statement and a little more impartial. Hair comparisons were done with the aid of a comparison microscope and required hours of examination. With the advent of DNA testing, the quality of hair examinations improved. The microscopic comparison became a preliminary exam instead of the final exam.

The brown hair seen earlier at the scene on Francke's scarf was about eight inches long. That eliminated anyone with shorter hair, provided the person hadn't had a haircut between the incident and the head hair standard collection. How can you be sure when weeks and months pass before the standards are collected? Microscopic examination revealed that Franke's wife could not be excluded, which seemed reasonable at the time. In his front shirt pocket, I found a stack of business cards, and to my surprise, there was a cut in the business cards. Apparently, there were at least two attempts to stab Mr. Francke as seem in Figure 37.

The cut in the business card looked like it was made with a knife with a thin blade. What could we learn from a cut in a stack of business cards?

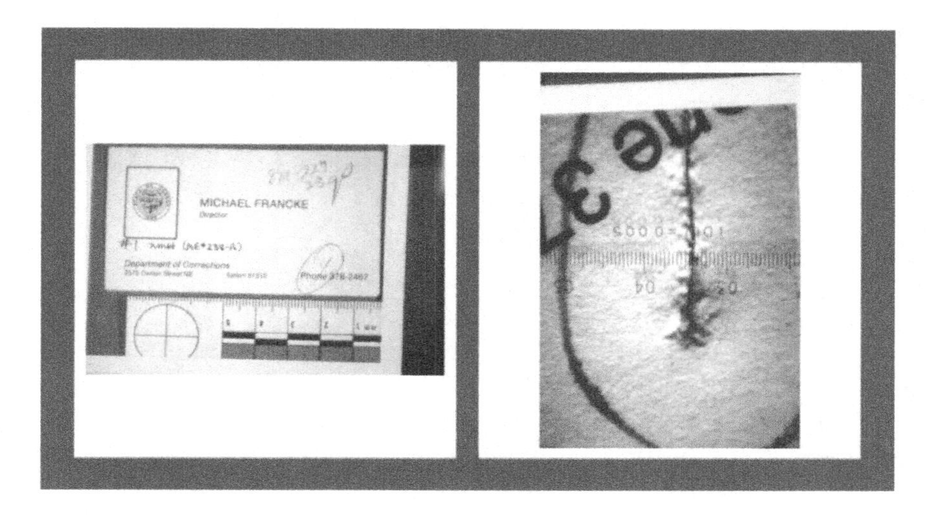

Figure 37

Our own business cards were of the same stock as the ones in his pocket, so I made a few attempts at stabbing them. Width of the knife blade did matter. Mike and I had a large collection of knives available to us that had been submitted to the lab for blood testing. Most of the knives submitted earlier were still in the evidence locker. We began stabbing business cards with various styles of knives to see what the cut would look like. From our observation of the cuts, there was value in this testing.

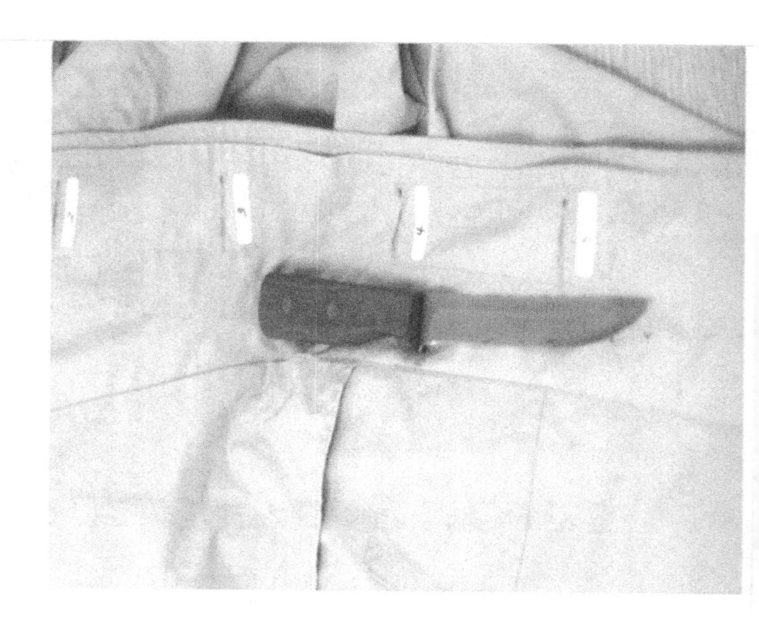

Figure 38

We laughed that after we were days into this testing, we had to be the foremost experts in knife cuts in business cards; what an achievement. Based on our testing, the cut in Francke's business cards was made by a knife with a thin blade. Two types of knives made a similar pattern; one was a fillet knife, and the other was a long thin-bladed knife common to butcher block sets.

The knife shown above in Figure 38 could have made the cut in the business card. It is resting on a white trench coat, and that could have been a source of concealment. Concealment is obtained on the button hole side of the coat; here threads are cut, and the knife will fit inside the fabric along the button holes, which acts as a sheath.

The knife is totally concealed from sight. In an investigator's meeting, we mentioned our findings on the type of knife used. One investigator had interviewed a woman whose husband had taken a similar knife from her butcher block.

He was living in the Salem area. He quickly became the center of attention and was later arrested for the crime.

In addition to the clothing and the knife issue, we still had to deal with the shoe impressions on the portico, and there were a lot of them. They all seemed to be an athletic shoe pattern, but dissimilarities did occur. The impressions were not in dust but were more permanent. The impressions looked unusual, but we just could not put our finger on what was wrong. Mike contacted the building maintenance people and asked who frequented the portico near the windows. The manager stated that the inmates from the prison did the window washing. Mike got a few pair of the shoes they normally issued, and we went back to the portico. The shoes were similar but the pattern looked like a reverse image of what was on the portico. Then Mike got an idea. First, he washed an area of the portico with a mop. He allowed it to dry a little and pressed the shoes into the dirty water. The shoes made an impression, but the impression was a negative of what you would expect to see. The dirt would accumulate in the spaces of the tread as the shoe was pressed. Normally one would see the most prominent portions of a tread in a shoe impression. It was amazing; we were getting a negative image of the shoes, the dirt accumulated in the spaces. This was something neither one of us had ever seen before. Looking back on the Alibi Tavern case, Mike and I found ourselves working on the edge of the known science available, a place where there are no simple answers.

We solved the issue of the shoe impressions in finding that these impressions had nothing to do with the Francke assault. The photos in Figure 40 depicts the unusual impressions and Mike doing his mop test to demonstrate how they were made.

Again, we were in a place where we had no references. This was new science. It was easy to see what was happening; it was just that no one had previously demonstrated this action. We needed to be careful and conservative.

Unusual Shoe Impressions on The Portico of the Corrections building

Figure 39

Figure 40

Mike and I were almost a year into working on this case, and trial was coming up. We had written almost forty reports. The number of clothing items that passed through the lab was probably over one thousand. The number of tests for blood, twice that number. The number of knives examined was approximately four hundred. The number of shoe-impression comparisons in soil was about fifty. We had hair standards from thirteen individuals for comparison to the hair on the scarf and vacuum sweepings from Francke's vehicle. There were several hairs in each vacuum sweeping bag, but if you only pick out one, the total number of comparisons were 221.

In reality, the total number of hair actually examined was well over five hundred. Then there were hairs from the clothing of other suspects besides their standards. Keep in mind, a standard would include several head hairs, pubic hairs, arm or chest hairs, and mustache hairs if applicable. Each comparison began with the examination of several hairs from the donor in order to become familiar with the variability within the hair standard. A hair standard may or may not represent hairs from all the regions of a person's head, but we knew that.

The question was if the unknown hairs could be eliminated as dissimilar to the submitted standard. For a method described as indirect evidence and junk science by some, we spent hundreds of hours on eliminations. We had no doubt the procedure had merit in some instances. If the suspect had red hair and you were given a brown hair unknown, the unknown was dissimilar to the standards submitted.

So far we had solved the shoe-print issues, located and identified possible knives, solved the shoe impression in the blood trail, and prepared a to-scale graphic of the portico. A suspect had been arrested, and the media was riding high, stating that law enforcement had the wrong guy, the police did not know what they were doing, and in no way would they get a conviction anyway. But we were not done yet.

The EMTs were interviewed, and they stated that Francke was on his back when they got there. How did he fall with the scarf lying neat and orderly on top? The DA wanted to demonstrate how this could happen, so we got Bruce, one of our lab associates who was about the same stature as Francke to be a model. We put a trench coat and scarf on him and proceeded to have him fall in a manner that would leave the scarf on top. We could not do it repeatedly. If it did not work here, how could we be sure this would work in the courtroom? After several attempts, our model was getting tired of falling and getting up again. Then it occurred to me: Place the scarf where you want it and have him get up. Do it the opposite. It worked.

The photographs in Figure 41 and 42 was taken during a rehearsal in the courtroom with Bruce wearing a coat with a black scarf. I have a wood paint stick in my hand representing a knife. We were also attempting to demonstrate how the attack occurred.

Figure 41

Figure 42

The prior photo was how we placed Bruce on the floor to represent Francke's position.

In the month prior to trial, Mike and I spent most of our time building court displays. We had a lot of evidence to show and wanted to make sure the jury would understand what we had to offer. A room was set aside in the courthouse that became our war room, and we proceeded to fill it with displays. One day I had a real knife in my hand while Mike and I were talking about the case, and I accidentally dropped it. The point struck Mike in the leg, and he immediately started bleeding. I thought it was funny, but Mike, not so much. He wanted to go to the ER. I liked the look of the cut compared to the what we saw at autopsy and in the business card, so I would not let him go to the hospital until I could find a camera and get a photo. Mike received a couple stitches for the incident, and I apologized profusely despite it being useful information. The single sharp edge could easily be seen. We eventually laughed together on what lengths we would go through to make a point. Or get poked with one.

I could just see myself giving testimony: "I know this to be true, because I stabbed my partner and the resulting wound looked the same." That is Mike's leg in Figure 43.

The next photo, Figure 44, is Mike in the war room with a display of the various knife experiments we conducted. He was keeping a safe distance from me.

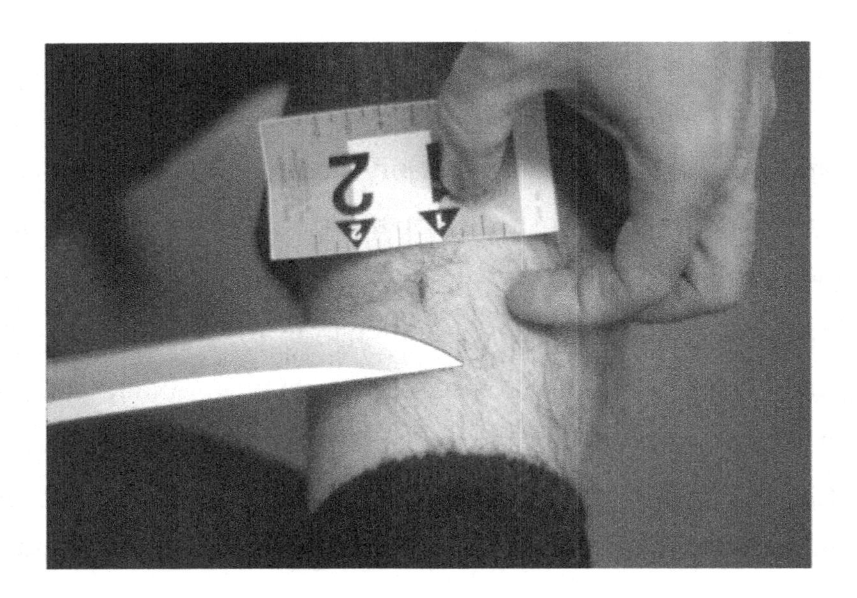

Figure 43

Figure 44

The trial itself was a media event. We had our war room, and the media had the entire lobby of the courthouse and several large trucks outside with satellite domes.

We were there nearly every day and were back and forth between the DA's office and the courthouse. The media continued its attack on law enforcement about the way that the investigation was botched. But we proceeded anyway. After the first day of my testimony, Mike and I were the center of the attention. We walked out of the courthouse, and the camera people were all around us. By the time I made it to the sidewalk, there was a camera man walking backward in front of me with his lens just beyond my nose. In a flash, he had a misstep and fell over backward with his camera. He leaped back up and was in my face again like nothing happened;

this was getting uncomfortable. We had to find another way out of the courthouse.

Inside the courthouse, things got weird. The judge took a break during my testimony, and I was still in the witness stand when the accused and his previous attorney stepped forward to be married. This was a 'wow' moment. I had done a lot of testimony, but this was a first. Apparently, the long hours of trial prep evolved into something more than just conversation. I had to wonder, what was next in this case?

For several days, Mike and I continued to testify about the evidence: the blood trail, the knives, the crime scene, hair comparisons, shoe impression comparisons, clothing exams and some blood typing. We did not have a single piece of evidence that directly linked the accused to the crime. Our role was to demonstrate we had diligently pursued the evidence and had eliminated others when we could. Several other witnesses testified, many of whom were street people who gave bits of information that supported the accused as being the one who stabbed Francke. The jury deliberated for quite a while, and to our surprise, the accused, Frank Gable, was convicted on all charges.

But it was not over! Twenty years went by, and I got a call from an old friend who was a very good detective with the Marion County sheriff's office, basically the Salem, Oregon area. He was getting on in years, too, and his conscience was bothering him about this case. Back in the day when the case was active, he interviewed a guy who had confessed to committing this crime. The guy had information that was not out for the public and seemed believable. Somewhere along the course of this case, the suspect information was sent to OSP, and reportedly this guy was cleared. My friend did not think this was right. Now there is an appeal in process. Perhaps the door will reopen, and we will do all of this all over again.

Chapter 18 The Parent Aid Program, a Crime Lab's Contribution to Families

As managers, the OSP headquarters often challenged us to open new doors in the fight against crime. There was seldom a response other than some new idea for traffic enforcement. To me, this was an opportunity. Law enforcement is in the business of after-the-fact management. The police do very little to deter crime and normally act after the crime has been committed. We all knew that the primary deterrent to crime was our presence, the presence of a patrol car, or the sound of a siren. If you wanted to actually do something about crime, you had to only look at the two primary foundations that lead to most crimes: domestic violence and drugs. I had already set up a successful program in domestic violence that turned out well, so what about drugs? How do you go about impacting young people before they get hooked on hard drugs and law enforcement becomes the last intervention through arrest? By then it is often too late for the young person.

I had high-school age children of my own and was also serving as chairman of the local school board. Was there something there that would work?

I was short on ideas, so I went to the junior high and visited with the principal about drug use in his school. He was aware of the problem, had seen the changes in the children, and had to deal with the parents when the grades dropped or the behavior got out of control. In meetings with the parents and the principal, the topic of possible drug use was mentioned at times, but it almost always got a quick response from the parents that that was not a possibility with "their" child. The interviews would end in a stalemate with unhappy parents and a frustrated principal. Even the suggestion of drug use caused anger. What I did observe, this interaction with the school may have been the first time someone outside the child's family noticed there may be a problem with the child.

From there, I went to the local medical clinic to see what they had available; I knew they did drug testing for employment purposes. They said they could help. But the cost of getting the urine was twenty-five dollars for someone to witness the urine collection, and the initial test was about sixty dollars more. Before any student could be tested, there was the issue of having a physician give the approval for the testing. It was getting complicated. I knew several of the physicians in the clinic, and they were woefully unprepared for dealing with drug use within the teenage community. They would never admit it, but that was my opinion.

There was an established pathway leading to intervention of a drug user. First you had to get caught by law enforcement, and then the door might open to a program by court order. The program would take the child, but the parents were often excluded until the program was complete. For instance, a drug abuse specialist could counsel the child, but under HIPPA rules, he or she could not discuss the child's case with the parents. If the parents wanted in, he or she would accept the parents separately for an additional fifty dollars per hour. This was never going to work. Without parents involved, I saw little hope here.

I called several state agencies to see what they were doing. As expected, everything was expensive, or the people who could help were understaffed and did not want to take on any more work. It might be different in other places, but in Coos County, everything was complex.

The principal told me that about 6 percent of the school population caused 90 percent of the problems at school. Of the 6 percent, 3 percent would pull out of it by simply growing up by the time they graduated. The other 3 percent became the responsibility of law enforcement. I have no idea as to the origin of that information other than the experience of the principal. No one knew the secret on how the other 3 percent pulled out of it by graduation, either.

Was it self-determination, or was it having parents who cared? Perhaps some of both. I'd had a troubled youth, and self-determination was my only way out.

I could see that if we were going to be successful at any program, it had to include the parents. What if we just provided the urine testing and gave the results to the parents? Could it be that simple? The crime lab division had a toxicology section and it would be simple enough if I could convince our own department to do the testing. The first responses were no. They did not know how many samples they would be talking about, and this was outside the mission of OSP. I reminded my chain of command of the directive to seek new ideas in law enforcement from the HQ memo. Division finally acquiesced to a pilot project.

Now I had the approval to do something, but how to start? I went to my staff, and we kicked the idea around until we came up with a plastic bag containing a sealable urine cup and instructions for collection and delivery to this lab. The only information we would request was the name of the parent, a phone number, and the county of residence. We would send the specimen out to the toxicology. lab, and they would report to us with the result within a week. Then we gave kits to local law enforcement agencies, and I took a few kits up to the schools. I contacted the local newspaper, got some good press, and waited to see if anyone would respond. It did not take long.

Late one afternoon, a dilapidated old sedan with a bad muffler pulled into our parking lot. Cheryl, the secretary, quickly advised me of the situation and suggested we call backup. The lab had drugs on hand submitted as evidence, and we could be a potential target. With phone in hand, we watched to see who got out of the car before we called for uniformed officers, and I locked the door. It was a woman with unkempt hair, and her clothing attire was modest, like she was someone from a low-income origin. She did not look like she meant to do us harm, so we waited for her at the front door.

After cautiously welcoming her, we asked what we could do for her. She said she had heard of our drug testing program. She had lost her son to drugs and now wanted to save her daughter. She did not have any other options available as she had no money. Could we help her? She had nowhere else to go. We quickly realized there was nothing else available quite like what we were offering at the time. I gave her a kit with instructions and said that if she would return the kit to us, we would do everything in our power to help her. When she walked out, my secretary and I were nearly in tears. We were new at this and determined not to let her down. If we could impact the poorest of the poor, what better way to serve the community?

The motto of the Oregon State Police was *To Protect and to Serve*. Under the OSP values statement were the words,

We will serve all people and fulfill our duties with the utmost understanding and empathy.

I was proud to reach out to the community unlike how the labs had ever done before. We were hopeful.

A few weeks later, I got a call from the junior high principal, and he was excited. He had had one of those dreaded parent meetings about a delinquent child, and the usual options were mentioned. The parents gave the usual denial response: “Not my child.” Out of frustration, he gave the parents one of our kits and told them to go find out for themselves. The test came back positive for marijuana, and the surprise of all surprises happened. The parents came back to the school and wanted to know what they and the school could do together to monitor their child. The principal wanted more kits. He was excited; we were excited. Word was getting out, and we had interested citizens come to our lab to volunteer in the program. They made kits and then went out to all the schools in the county and pushed the program’s early successes.

When we were a few months into this, we were getting specimens daily. I was surprised that the parents would put trust in us as a law enforcement agency in dealing with their child, but they did. After receiving a phone call on the results, they would call or stop by to see me about options.

I had to brush up on local agencies that dealt with drug abuse so that I could refer the parents to them.

I Learned quickly from what parents were telling me what worked for them, and I was passing the information on to other parents when asked.

The most common problem I would hear was about the rights of the children. Somehow folks had the idea that their child's room was off-limits to them based on the child's constitutional rights. I would tell them the constitutional rights applies to governmental agencies, not to their children in their home. The parents were not running a household democracy; no one took votes, and majority opinion did not rule. It was their home and their child, and they had a right to go anywhere and search for contraband. What an eye-opener for some parents.

In my discussions with the parents, I said that the best model was for the parents to introduce the kit to the child, place it on top of the refrigerator, and state that anytime they were suspicious, they would require a specimen. If they requested a specimen and the child refused, then the child was grounded, no car keys, or other punishment was the result.

In the program design, the state police would allow specimens only from the local county in order to keep the numbers down in the pilot project.

But word was getting out, and parents were showing up from other counties with sad stories of problems with their children and nowhere to turn. Despite the rules from HQ, the staff and I talked it over and decided that we were going to fudge a little bit. No one would be left behind, and we proceeded as if they came from within the county. How could we possibly turn anyone away? It was unthinkable.

At the end of the first year, I needed some feedback on how we were doing and where to improve. I held a meeting of school representatives, law enforcement, and parents who had used the program and were willing to talk about their experiences. The parents gave remarkable testimony on how it was working for them, and the teachers were equally positive about having an option for the schools. It was unprecedented that a parent would come back to the school after one of those dreaded meetings I mentioned earlier to seek school cooperation with their child. That action was probably more mind blowing than anything else. Now it was happening with regularity.

The first year the OSP included our program in their budget requests and received positive responses from the legislature in funding. The legislature knew about us in our tiny lab on the coast. Everyone was happy. Now other lab directors wanted to push the program in other areas of the state, but to my surprise, the OSP said no. In one case, the lab personnel stated they would come in on their own time to prepare samples and get them sent off. It was great marketing for the crime labs.

Why not proceed? Suddenly it became apparent that the OSP was not interested in developing the program statewide. They wanted to quietly discontinue it in my lab, too.

My staff and I could not believe this was happening. Apparently, the cost of the program was an unknown. Then it happened, my division captain drove to my lab to stop the program. I was waiting for him.

I arranged for a room full of parents, and we had a meeting. They gave testimony on how the program had impacted their families, and some even stated that they believed it saved their child's life. Tears flowed for over an hour. Unlike the lady in the old car from the beginning of the program, one well-dressed man provided the most impact. He appeared well-to-do and had a teenage daughter who was a problem for him. She was into drugs, running away at times, and doing poorly in school. He had had her in several treatment programs, but eventually she would be home again and back to old habits. Therein was the problem: controlling her behavior from the home. Until this program, there was no way of knowing if she was back on drugs. Placing the kit on the refrigerator and doing random testing brought her in line.

Now her grades were improving, and she was responding to her mom and dad with love again. He was positive that the program gave him his daughter back. I talked about when my own daughter came home from school one day to report that she had overheard two other girls talking at the locker next to hers. One girl was trying to talk the other one into going to a party that night. The other one stated that she could not go, as her parents had one of those kits and was going to test her if she did anything they thought was suspicious. Bingo! We were there. What was the captain going to tell these parents now?

The meeting concluded with the promise from the captain that the program would continue for a while longer. He knew he had been bushwacked and was not happy, but I was desperate. I took my scolding, and we moved on. I could not understand how the state police would look negatively on a program that did so much good. But they were afraid that the testing would really take off and become a financial burden on the state. I can't blame them for that; it probably would have, but I believe it would have been money well

spent. But my opinion did not count for much here. As I said before, law enforcement normally acted after a crime had been committed.
Some folks just could not conceptualize it any other way.

One morning I got a call from a reporter back east who inquired about our program. After I explained what we were doing, he asked if I would be willing to be interviewed by Barbara Walters on *Good Morning America*. Wow! That was exciting, and the answer was yes. He told me that I would have to go to a television station in Eugene for the interview, which would occur about three o'clock the following morning. I advised HQ, drove that night to Eugene, and went to the station. There was no one there but a camera man and me. I sat on a stool in front of a camera in a studio for over an hour before it was my turn. I could hear and see Barbara Walters on a small monitor, but I had to respond to her questions by talking to the camera, a large and inanimate object. It was odd and uncomfortable for me, but I managed to get through it. She concluded that the program was "probably" all right, as long as the results were controlled by the parents and there were no reprisals by the agency toward the child.

Kits and parents continued to come into the lab, and it was time to do a review of how we were doing. First, I compiled a list of drugs that were detected in children through the Parent Aid program. The percent positive was about 50 percent, and the primary drug detected was marijuana. There were a few hits on cocaine and a few prescription drugs, but marijuana (MJ) was the primary drug detected in better than 90 percent of the positive urine samples. MJ is in a class called a *gateway drug*. This is where the experimentation usually begins and may lead up to hard drugs later on in the experimentation.

I obtained results from the toxicology lab on what percent of all specimens outside our study were positive and how the drugs detected were distributed. Most of the urine samples submitted by law enforcement were positive for a drug and, in some cases, multiple drugs. Keep in mind the samples were collected by law enforcement due to a crime related stop. The most common drug was methamphetamine, followed by MJ and the opiates. It was clear that the Parent Aid program was impacting the youth earlier in their drug use, with MJ as the primary drug and almost no methamphetamine detected. Over 90% of our positives were for a gateway drug. Meth was considered a hard drug and epidemic in some communities. Parents were doing a better job than law enforcement at getting to their children sooner in their experimentation.

I published the results of my comparison to the department but did not get a response, a bad sign. I had received inquiries nationwide on the program, and I sent the report to several locations across America. It was clear that no one anywhere was producing the successful results that we were getting. The parent partnership with the schools was a big factor that no one previously had bridged. I received accolades from everywhere but OSP. The Oregon Peace Officers Association awarded me with an outstanding service award for developing the program. At the presentation meeting, my division supervisor in HQ hardly acknowledged my presence. There were no congratulatory handshakes from him after the ceremony. My associates, Cheryl Waddington and Gail Sampson and I were undeterred, we continued on with the program.

The program continued for a few more years until HQ finally pulled the plug. There was nothing I could do. I received calls from several agencies and a single police department with a forward-thinking chief picked up the program and offered it to parents in his community. For our community, the service was gone.

To do the testing, he used a special UA cup that had the chemical drug testing indicators attached to the side and could tell you if some of the basic drugs were present. He also offered personal help to the parents, similar to what we had done.

The one downside to his new design was that the UA was collected at the station with a uniformed officer handling the case. This was uncomfortable for some parents, as they thought if the test was positive, the police might target their son or daughter. The program still moved forward for the agency. Eventually the Oregon Chiefs of Police Association picked up the program and developed it for other police agencies in the state. Currently there are more than fifty agencies in Oregon offering the program to their communities. Recently I was told that the state of Oklahoma has decided to pick up the program and was implementing it through their law enforcement agencies.

I never saw the woman in the old car again. But we often wondered about her and her quest to save her child. I hoped this mother had found a way to help her daughter somewhere. She will never know how much she empowered us to make this program a success.

This program and the domestic violence program that I mentioned earlier were personally the two most rewarding contributions I made to law enforcement during my career in forensic science. With both programs, all of us in the lab believed we were making a difference. It felt good to go to work in the morning.

Chapter 19 Retirement and Another Beginning

A career with the state was a good job and provided my family with the essential needs as well as keeping me excited about my work in the lab. I ran a small crime lab, and as the director, I had the responsibility of doing casework as well. This suited me perfectly. As the years went by and it was getting close to retirement, my wife and I would seek counseling on the potential for retirement. The funds were there, due to good planning, good investments, and a retirement plan from the state. The common statement was to keep working as long as you like and, when you have three bad days in a row, leave. At the retirement age of fifty, it's amazing how your attitude changes when that big day comes around. You spend your career looking forward to retirement. But when that day comes, you keep working because you know you can go at any time, so that pressure is off.

I suppose if I'd had a job that I did not like, it would have been different, but I kept working for another four years and would have continued had conditions stayed the same.

The department decided to put a person in charge of the labs who had no lab experience at all.

He had floated around the department and was a problem wherever he went. In the infinite wisdom of someone in upper management, they moved him to our division.

When we had director's meetings, he did not understand what we did. He was a numbers man. What he did understand was that there had to be punishment and accountability that was palpable. The other managers were not interested in taking a highly successful system and heading into uncharted directions on the whim of a person most of us deemed unqualified to be a reagent tech. I think there was a holdover in the idea that we were not real men but technicians from his viewpoint. The conflict continued for several months, and then some of the directors were suddenly under investigation for unspecified reasons. During management meetings, we were referred to as the *old dogs*, or other such words, implying that we were too old to understand where we needed to go.

Some good people were put on administrative leave; others were demoted or forced out. Keeping a leadership position became a matter of survival against the dumb and dumber. What was interesting was that the core group of management, which had been in place for twenty years, left. When the new management positions became available, no one was willing to step up and apply for the lab director positions. These were high-paying positions, and no one would apply? That should have been a red flag right there.

The new leader made the profound decision that you did not need a science degree to run a crime lab, so he opened the positions up to anyone with management experience in anything. Several people applied, and I was on the hiring committee for some of the new positions. It was sad. I had to interview a manager from the local Taco Bell for a position as a laboratory director. This was incredible but reality. We just shook our heads. How bad were things going to get? We were about to find out.

In the upper management, no one was listening to us. Experienced laboratory personnel were leaving, and those remaining refused to apply for advancement. The non-management scientists were protected under the union, and it was safer there than to consider the additional pay. Besides they got overtime, and management did not. At the end of the year, most of the scientists who did crime scenes were making more than management anyway. Eventually the division director befriended a few people within the division, and they were promoted. This was not based on qualification but on who was willing.

At least once, they did look outside, and one individual was promoted to the position of assistant division director who was from a university background and held a PhD in chemistry. Maybe things were looking up. The new assistant director made a tour of the labs and returned to Salem with a whole host of good ideas for solving problems within the division. That was not met with approval as there was no discipline involved. His job was to go out and get information and the names of persons requiring punishment. He failed at this, and an argument followed. The new assistant division director was gone a few days later.

It was not long before there was a replacement. This person was a forensic scientist but not someone you would call ready for the position or someone the rest of us would have supported. He was there because he was willing. His title was now quality-assurance officer, and he set up some QA tests for the rest of us. It was not long before he sent me and another director a QA sexual assault kit. We both worked it up and submitted the results. But to my surprise, I missed it, and so did the other director. We compared notes and decided that we were not wrong. We requested a meeting before the results became part of our personnel file.

It was quickly apparent that both of us knew much more about forensic serology than the new QA officer. We explained to him why we were right and he was wrong. He eventually acknowledged that we were right, but he said he was not going to change the result of the QA in our personal file.

What? If you want to mess with the mind of a scientist, be irrational. Apparently, it was a new plan to attack the working folks.

Gary, who was one of the longtime lab directors and one of our very best people, was next to fall into disfavor. I was called to Salem and advised that I was to transfer to Medford to run his lab. I declined and said I could run it from where I lived until a new director was hired. Upper management agreed. When I left Salem, I called Gary and told him what was up. He was unaware of what was happening and decided to retire instead, and he left shortly thereafter. Gary was one of the best bloodstain pattern experts in the division. Now I was running two labs.

The division director was interested in the number of cases a person could get out the door and not in the quality of the work in those cases. I was directing two crime labs at once, and an example situation came up in the fingerprint section of the other lab. There were several people in the section with varying degrees of skill and some specialization within the individuals. All of them were highly qualified and did their own division of labor within the section based on what was expedient. The newer people were assigned the cases with limited evidence and got a good number of cases out the door that had few or no prints.

The more experienced people worked the complex cases, often with large quantities of evidence. The section was working well. The case volume was good, and they were getting it out the door with a quality product. The backlog was low. As is common, when there is a problem to solve, often the people closest to the problem are most qualified to solve the problem. The people within the section were working to their individual strengths and getting the work done.

The problem perceived by division was that some individuals were not pulling their load when you looked at the stats based on the number of cases going out the door. The solution was a demand for discipline at the hands of the director, me. No amount of explanation seemed to make any difference. In my opinion, if all labs receive a similar caseload over a six-month period, then the total output of the section should be evaluated on a division-wide basis over that time, not on an individual basis, but that was not the rule of the day.

I had one individual working on a major burglary case that took her a month to complete. She generated several prints and a subsequent match to a known burglar, and an arrest was affected. This was a serial burglar, and we were all proud of her work. The agency involved wrote a nice letter of appreciation to the lab. Within the division, I got a call again that this person did not meet the statistical requirements assigned to her specialty. If I could not fix it, they would find someone who could. So I went back to the section, called a meeting, explained the situation, and told the examiners to make sure that the outcome of their monthly workload was balanced and to do whatever it took. This worked for a while, but it only made the entire section fall behind when a big case came through the door. Now the section was not keeping up.

Then things got worse. We started losing the good people, and the numbers game was killing us. Instead of tackling the large cases, we were told to screen them, work the few essential pieces of evidence, and send the rest back. Consider for a moment if your home was burglarized, ten objects had been moved by the burglars, and you just knew their prints were somewhere on those items. Those items were sent to the lab for prints, but only one item would be selected. How would you know which item to choose? And if no prints were located on that one item, was the case over? Apparently so.

The need for numbers was damaging the reputation of the division, and the community we served was the final loser. Police agencies were upset at the lack of testing and the length of time it took to get the results. It was interesting to see how one change in management policy could change the entire work product and put it into failure. But that did not seem to matter. It only showed that the lab management was not doing their job and that the division director needed to apply more pressure. Remember the old saying, *The flogging will continue until morale improves*?

Over the years, we lab directors were well trained on managing employees and getting productivity through encouragement. We used discipline as a management tool only as a last resort. Now discipline was the rule of the day, and other tools were of no consequence.

As I watched my fellow directors go down in the dust or retire, I knew it would be only a matter of time before my number was up. In the meantime, Mike, a good friend and fellow director, was promoted without a pay increase to a supervisory position over the rest of the directors but still under the captain. The state system at the time had eight crime labs.

One day we both received a letter from the division director stating that I was under investigation for criticizing the agency outside the confines of the department and making a series of untruthful statements about the division director, him. The document named persons, statements, and places. Since I was the target, I was advised not to interfere with the investigation. That meant that I could not contact the persons mentioned in the complaint. Mike was told to do the investigation and, if it was true, to fire me. If he could not fire me, he would be replaced. A catch-22 in real life, as the captain did not like either one of us. By this action, the division director was going to get one us in his plan.

We both took this very hard and were without an immediate solution. We both had great careers, and now it was down to this? Faced with the consequences, I took matters into my own hands. Contrary to what I was told, I did my own investigation and found that the statements reported by others were not true as stated in the complaint. The people involved never made those statements to him. In fact as much as I disliked the guy, I was always careful to whom I vented my feelings. The problem was that I could not tell anyone about what I knew. I had to wait for the investigation to discover it.

With this complaint and other previous problems, Mike was having medical problems over the stress and ended up in the hospital for a week. I was not doing much better. We had devoted our hearts and minds to this career, and now it was as if it had all been for nothing. It reminded me of totalitarianism. You simply get rid of anyone who has the capacity to think and promote those who are not capable of replacing or questioning you. It got to the point where I could no longer function in the lab. I wandered about and could not focus on my work. As the days progressed, I became less productive.

Finally, my employees, not knowing what was happening, came into my office one day and held an impromptu meeting on my behalf. It is good when you have staff like that but embarrassing to know you are slowly losing your mind. I opened up and laid it all out there on what was happening. I was losing my mind from the stress. They could see it, too. Up to this point, I thought I was still in control of my feelings and my job. I was not. The tears flowed, and I went home and contemplated my future.

This could not go on. I had a very successful career, a twelve-page résumé, and more awards than anyone in the history of the OSP Forensic Service going all the way back to 1939. I'd had lots of high-profile cases and successful prosecutions, written several research papers, and now was destined for a dishonorable discharge. I was at a low point in my life, and it was time to do something, to step up to the plate, as I often told my children in hard times. They all knew what Dad said in times of crisis: "When the going gets tough, the tough get going." I needed to pick myself up by my bootstraps and listen to my own advice.

I knew that the superintendent of state police had an open-door policy that stated that anyone in the department was welcome to come in if there was a problem that could not be resolved at a lower level. I figured there was risk, but I was going to lose my job anyway,

and I was not going down without a fight. The next day, I drove to Salem, found the deputy superintendent in his office, walked in, and laid it all out on the line.

I had known and worked with him when he was a detective many years prior, and we had always been friends. He listened to me and decided to call for a management audit of the entire forensic division. These audits are done by people outside the division who report directly to the superintendent. I was partially relieved. He said that I would need to cooperate with the audit but that I would be protected during that time.

Some of the other supervisors were now ready to list their complaints about the division director when they heard about the audit. But the investigation on me was still there, and either Mike or I was going down. In the meantime, one other director was accused of taking a firearm from the lab for personal use. He denied the allegation but was sent home on administrative leave anyway. These investigations always take a long time. While he was on administrative leave, his retirement date arrived, and he left without so much as a retirement coffee. Twenty-five years and not even an official good-bye. The investigation finally revealed that the charges were unfounded. But the damage was done, and another lab director was gone. Besides a director, he was the division's foremost firearms authority.

On my way back from Salem, I stopped in Eugene at the Bureau of Labor and Statistics and filed an age discrimination complaint against the Oregon State Police, the second step in my plan. I knew it took about a year for one of these complaints to be investigated, and by law, during that time, the department could not touch me. Call it extra insurance.

I called Mike, and we were both relieved. There was no need for his report as there was no reason to act at this time. While the state of Oregon reviewed my allegations through the audit, I was not to put myself in harm's way with the department and was to maintain a low profile. Apparently, the superintendent of state police was quite unhappy with this new development and called the division director into his office for a chat. We were not privy to the discussion, but rumor had it that the talk was not positive reinforcement for the division director. He was instructed that he could not touch me or do anything that might jeopardize the state's investigation. Reprisal was not an option. Now it was a waiting game. Could I outlast this guy? At least I had the time to think about what I wanted to do.

We had supervisor meetings about every ninety days, and I would be required to attend. I knew that I was insulated from discipline, so I would sit and listen but not contribute to the conversations. I knew this pissed off the division director, but I loved it. I was careful in that I would not contradict him or act in any manner that could be construed as obstructive. If the situation escalated, I could walk out the door and retire as a backup plan. In the days that followed, the other directors and I would converse and cover each other's back when we could. Despite all the turmoil in headquarters, the rest of us were keeping the division afloat and getting the cases out the door.

During this year of investigations, for some, it got even worse. In one of the labs I was running, a new circumstance presented itself. Whenever a report is generated in any of the disciplines, it is technically reviewed to ensure that the notes reflect the proper result and administratively reviewed to ensure that the entire packet of information was complete. One of the people doing the technical reviews was so pressured to perform and to meet the case quota that he would rush through the technical sign-offs to meet his obligations. Under enormous stress, he finally got to the point that he just signed the tech reviews and got them out the door without reviewing them. This was later discovered by the QA officer when a mistake was found. An internal investigation was started without my involvement, straight out of headquarters. The examiner confessed to the complaint and gave reasonable explanations for this occurrence. He was disciplined, but it did not stop there.

It was construed to be a falsification of public documents, and the DA's office was notified for possible prosecution of the examiner.

A career examiner now facing criminal charges? In theory, each document was a potential felony, and there were a lot of documents. There were no words to describe how we managers felt about this. Occasionally mistakes happen in all work environments; corrective action is taken, and everyone moves on. But criminal prosecution was a new level generated by a dangerous mind. I'm not saying how it happened, but someone contacted the DA's office and explained what was going on. The DA's office refused to accept the case, much to the disgust of the division.

When problems occur, managers are trained to first look inward at the system. What is it with the system that caused this person to act in a certain manner? Were the workload requirements too high for one person? It was rare for someone to make mistakes in our business with intentional malice. Was the system ultimately responsible? Normally after you have examined the circumstance of a misdeed and the employee took some of the responsibility, a progressive form of discipline was applied. It could be just counseling in the beginning. Now we jumped to criminal prosecution? The workplace into which I was hired looked nothing like what was happening now.

W. Edwards Demming was an American engineer and management consultant who wrote several books on how to manage people. In our management training, we had studied his ideas and especially his fourteen points to better management. Demming writes that when things go wrong, 94 percent of the time, it is management's fault. Failure to train properly, inadequate communications and specifically, don't rely on statistics to manage, manage through leadership. In his famous fourteen points, number eight is especially important and missing in the current management style: "Drive out fear from the work environment. Punitive management, where the primary check on performance is to punish performance that is not up to the expected standards, is counterproductive."

Finally, the internal investigation on me was dropped, the age-discrimination complaint was in the works, and the audit was finally complete. We were not sure how the audit came out for the division director, but we remaining supervisors who had complained were called to Salem for individual interviews. A major, the division director, and a detective were present for the interviews. Each supervisor was asked if he/she would support the division director, and if not, the supervisor would be demoted or potentially fired on the spot. It was unbelievable, again. All the interviewees fell on their swords, as they needed the job.

As each person came out of the interview, we offered support for that abused person. Seeing my friends seriously shaken by the experience, I thought it was curious that I was last.

When I went in, it was made clear that I was the source of all the problems and my career was about to end. After listening to what a bad person I was, I asked if the three of them had read the fine print on the request for an internal audit. Those who came forward were protected under a form of the Whistleblower Protection Act. This may have been baloney for all I knew, but I had to respond with something quickly. I felt the gotcha moment, as the three were silent and looking at each other. I followed by asking if the deputy superintendent was aware of this meeting and the content of the questioning. Then I went on to mention his statement to me about being protected.

Another moment of silence. Their response was confusion and backtracking as I stared them down.

When I walked out of the interview, I was unharmed and wished that I had been the first instead of last in the cycle. I am sure that they planned to make a final statement by shoving me out the door. Looking back, I wonder if they ever realized I was full of it when I told them I was protected under the Whistleblower Protection Act. Or was I? I had no idea.

I planned that the next time I went to confession, I would bring it up and seek forgiveness whether I needed it or not.

Driving home, I was thinking that this was about the last of this stuff that I could stand. I had been successful, but at what cost? My mental state over the last few months was at a point where my doctor kept trying to get me on medications. It was probably a mistake that I refused. I had learned a long time before that when you are in the middle of bad situation, you are often the last to see it. Looking back, that was me. The workplace had become a place of constant fear of what might happen next. The age-discrimination complaint was finally complete, and my objections were justified. The vindication did not drive me out, but everything leading up to it demoralized me to the point of leaving voluntarily.

My friends within the system were leaving, and the young replacements were trained under a new philosophy that they were to run tests, not solve crime.

My wife and I looked at our investment portfolio, and it looked good enough to go. In 2002, I pulled the plug and left. I had exceeded the three bad days in a row by quite a bit.

The division director eventually fell into such disfavor with the upper administration that one day he emptied his desk and left. It was three months after I retired—too late for several very good people.

But the absence of experienced people simply led to continued bad practices. One of my former employees, who was probably the best crime-scene reconstructionist remaining in the department, wrote up an extensive, brilliant report on a shooting homicide. He had done a lot of work, researched the data carefully, and it was an example of how an accomplished examiner could do a report. When the report got to the new management, there was a discussion, and it was deemed that no one else on the department could write a report like that and that it put unreasonable expectations on the other employees. He was counseled and told that his report was old-fashioned, and he needed to rewrite it, leaving out the reconstruction and presenting only his observations of the scene. The system had moved to setting the standard at the lowest common denominator. This experienced crime scene scientist transferred into drug testing and never went out on another scene. What a loss.

In the days that followed retirement, it was like a huge weight was removed from my chest, and I was busy catching up on honey-do lists or fishing. One morning after my wife had gone to work,

I went into the kitchen, looked around, and decided that with my skills I could reorganize this place and make it much more efficient.

When my wife came home from work, I was anxious to show off my improvements. She looked around, and her only statement was that it was time for me to go back to work. I think placing labels was the culprit.

Over the years I had developed a friendship with several prominent defense attorneys in the state, as we were often on opposing sides in high-profile cases. Being on the opposing side does not necessarily make you an enemy. It was not uncommon for a criminal defense attorney to call me about a case and want to talk about what I had or what it meant. At times my testimony had helped the defense case. One attorney, Steve Krasik, who was on the Francke case, contacted me about doing defense work. He said that his group had been waiting for me to retire. They knew that I was honest and knew that what I said would be truthful. He also said that when they had a case I had worked, they had known there was no sense in attacking me; they would not win.

Very flattering statements. I knew he wanted something, but I liked it. He asked if I wanted to go to work for the defense, and I said that I had no idea how to start, how much to charge, or how to even bill the state indigent fund for my services. He just wanted me to look at what the state had done and see if it was OK. Easy enough. He would help me solve the business issues.

I set up a business and developed a business contract. It was kind of exciting, and I was back into doing forensic science without the influence of people I did not like. The case was easy, and there were no uncommon errors or emissions that I could find. I got paid, and it was very gratifying.

My wife and I discussed the business situation and decided that if I could make a couple grand a month, that would really offset my retirement. Word got out quickly that I was available, and the phone was ringing several times a day. What I had hoped to make in a month was turning into daily income. I started buying equipment and gathering supplies in preparation for more evidence examinations.

A pivotal moment in my business career arrived when I was asked to assist on a homicide case in the Portland area. A vehicle was processed by the lab, and I had the reports and photographs. I requested to see the vehicle myself and went to Portland for the search. First, I found that the vehicle had been out in a lot with the windows down for over a year, with no attempt to preserve it for trial or the defense. The lab search at the time of the incident had detected blood on the driver's seat and no other places.

When I looked in, I could see blood on the ceiling above the driver's position. I knew a common mistake in processing a vehicle is failure to look up. I took photos of the blood and showed it to the defense attorney. When we were in trial, the scientist who processed the vehicle was shown my photographs while she was on the stand.

She admitted that she had not seen it and began to cry on the stand. By the time she got back to the lab, I was the worst SOB that ever lived. How dare I question what she did and then find something and report it to the defense? They felt the proper thing to do was to report it to the lab, not the defense. That was about how much we knew about defense work when we worked for the state.

Whatever friends I had in the labs when I retired were now pretty much gone. I had committed treason in their eyes. Most of them did not understand the law. Oregon law states that the evidence must be maintained in the event that the defense wants to examine it. The accused has the right to challenge the accusers. Also as a defense expert, I was under attorney-client privilege; I could not talk to them about what I found without the consent of the client or his attorney, and that was not going to happen. The prosecution also had the right to question me about my findings, and they did not do that. That was not my fault.

I was suddenly in a whole new world of forensic science. Much of it was quite different from my previous work, and I had a whole new set of laws to learn. Most of the work was reading reports, interpreting the results, and very rarely actually looking at evidence. My job with the state was 95 percent looking at evidence.

With the new role, I might be in court to watch the state's witnesses testify and pass notes on to the attorney. What a change in job descriptions.

It got so busy that I needed help. I knew the good folks who had left the lab and contacted several of them, but they were not interested. My friend Gary was teaching week-long classes on forensics for police officers at thirty dollars per hour. I convinced him to join me and help with the casework. What I needed besides his knowledge and peer-review capability was for him to develop as a graphic artist. He needed to learn Poser software. I purchased the software for him, and he became quite proficient at using it to depict crime scene circumstances that we would develop. He now is probably the best graphic artist in the country when it comes to crime scene graphics. Where else can you find someone to whom you can hand over the reports and photographs and tell what is on your mind for what happened and expect that within a few days renderings will start showing up on your computer?

We made a good team. Besides state indigent cases, we were picking up federal cases and military cases, as well. I was living in motels, hotels, and commercial aircraft for weeks on end. Years later I would incorporate as International Forensic Experts and have eleven other scientists within the business. Life was still good.

Chapter 20 The Allen Reavley Case, a Cold Case Gone Wrong

In Great Falls, Montana, in 1964, a large grocery store was managed by a man with a large family. Mr. Arrotta was the proud father of seven children, all small and the youngest still an infant. One night about midnight, several people came to his house. During an interview days later, one child stated that she saw a person outside her bedroom door but did not recognize him. The strangers were criminals and were looking to do a robbery. Both the father and the mother were forced to leave their children behind and were escorted to the grocery store, open the door, and enter. A witness saw several people enter the store that night but thought nothing of it at the time and later had no other specific information to provide. Earlier in the evening of the incident, neighbors remember hearing a vehicle in the alley behind their home, but no description of a vehicle was ever given.

The store manager was probably asked to open the safe. However, there was a problem; the store owner had recently installed a new safe with a timer.

It could not be opened at night. The next morning when the delivery people arrived, the door was unlocked, and they found both the store owner and the wife dead on the floor in the back room. Both had been stabbed to death.

The local police responded. Some photographs were taken, and a knife, Figure 45, was recovered from the floor near the wife. It was the same knife used for cutting the vegetables in the store and had a long blade and no hilt. The hilt was the part of the handle that prevented the hand from sliding down onto the sharp blade. Compared to today's standards, very little evidence was preserved. One witness stated there was an apparent blood smear on the entry door but no sample was collected. The store needed to open for business so the crime scene processing was reportedly rushed.

Later when I did an examination at the police department, I discovered that the tip of the knife was bent. As soon as I saw the bent tip, I knew how that happened. It was important that the murder weapon originated at the scene and was not taken to the scene.

Photos demonstrated there were pools of blood around both bodies plus a blood trail leading to the cash register at the front of the store. There was no evidence that the bodies had been dragged to this location from somewhere else. However, there were bloodstains and possible bandage material, (Figure 46) in the hallway between the back room and the main store. If the blood was from the victims, why were there bandages on the floor?

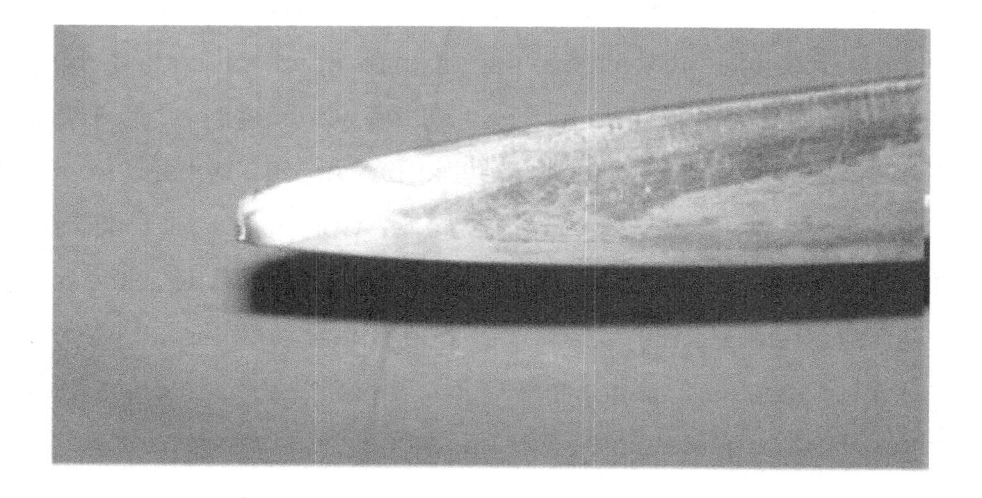

Figure 45

Besides the blood, rope used to bind the victims could be seen. One investigator stated that there were shoe prints in some of the bloodstains, but there was no blood on the bottom of either victim's shoes. Nor were there any quality photographs of the shoe impressions. There were a couple of possibilities for this portion of the scene. One, the manager may have been struck on the head at this location and tied up.

But the pool of blood seen in Figure 46 near the white cloth suggested more bleeding than a blow to his head. In my opinion, the manager was stabbed nine times in the back, most likely in the final position in which he was found.

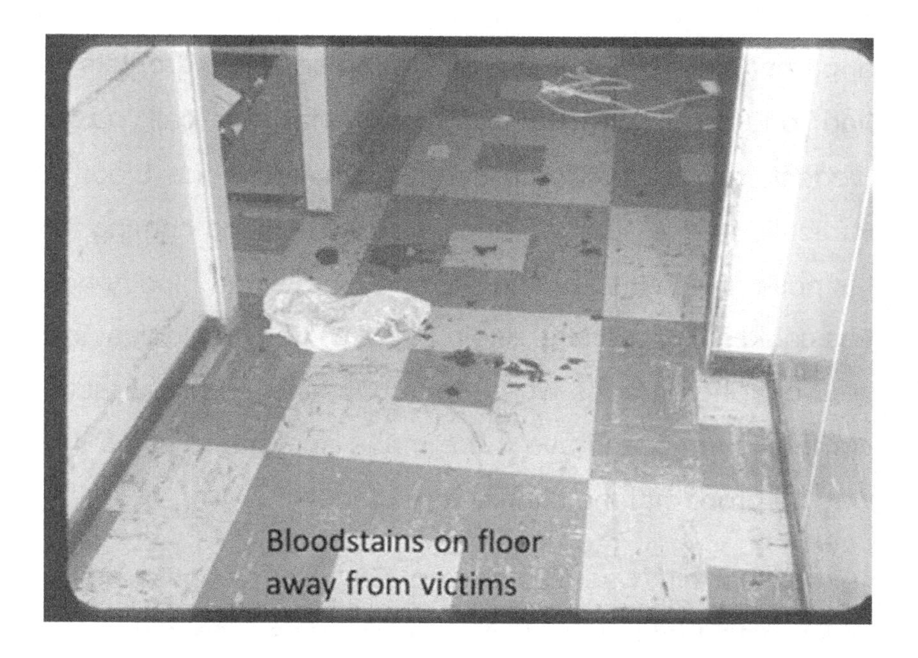

Figure 46

There was some bloodspatter near his head, either from being struck or coughed blood. Coughing blood is not uncommon when the lungs have been compromised and blood leaks into the trachea.

He also appeared to have bled from the nose, adding support for blood in the airway. No one found a blood trail leading to either victim, suggesting they were moved. Besides blood running down the back of his shirt, there were a number of small spattered stains as well.

These did not originate from his mouth or nose and might be associated with the stabbing action. Beside him were several apparent bloodstains that appeared to be single droplets that had originated directly above in what we call *passive bloodstains*; the only energy was gravity as the blood fell vertically from a source. Since there was blood coming from his nose, if he was standing at one time with a bloody nose, this could have been the origin or not. When you look at forty-year-old photographs where there was limited blood testing, some assumptions have to be made and stated since there was no proof that the stains were blood.

Near him was also broken glass from a bottle, none of which was under him. Broken glass was also against the manager's wife but not under her, either. The appearance of this glass was indicative of the bottles being broken while both victims were in the prone position.

The autopsy report on the wife stated that she had two blunt-force injuries to the head and was stabbed thirteen times from two different angles.

Several fingerprints were found, but no matches were ever made to anyone during subsequent investigations, including our client. The following statement was included in the FBI analysis of one of the prints in blood:

> *March 14, 1994*
> *Fingerprint 2-A is a fingerprint in the victim's blood (Type A) which was located near the safe and has been compared to the victim's and is not theirs. So this would be the killer's fingerprint.*

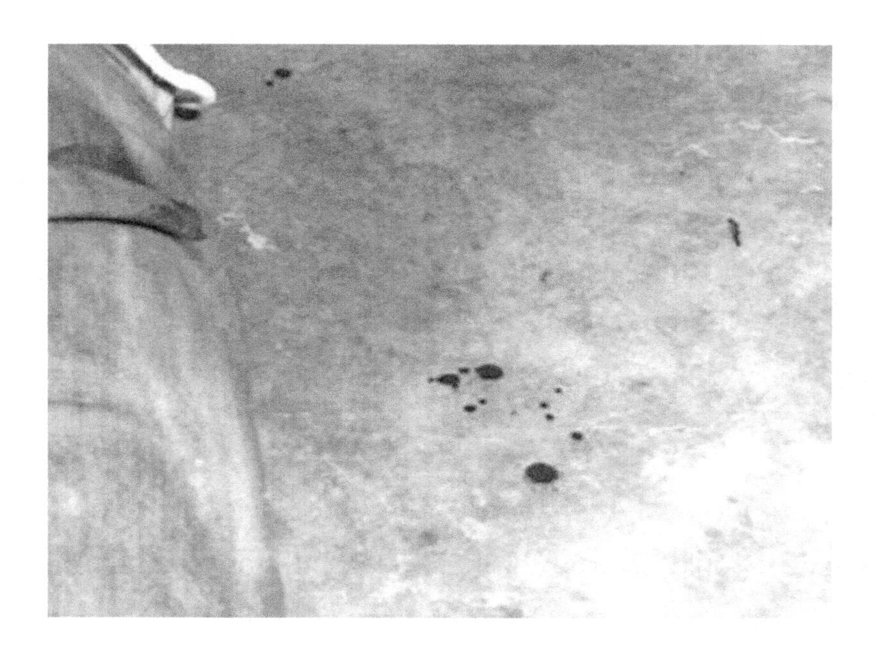

Figure 47

In the photo above, Figure 47, spatter from directly above the floor and next to the body is visible.

An extensive investigation ensued, but no one was apprehended. Numerous people were interviewed as possible suspects, and their reports were still in the case file.

The investigation continued for over a year with no success. Among the employees of the store was a seventeen-year-old box boy named Allen Reavley.

Back in 1964, Allen had taken some money from the cash register just prior to this incident and had been caught by the store owner and fired. For obvious reasons, the police went to Allen's house and interviewed him about the theft and his location during the murders. Allen's parents stated that Allen had been out for a short time that evening but returned and was home at the time the murders occurred, which was around midnight. Allen and his father had been out selling insurance that morning, and Allen's demeanor was not suspicious to the detectives. As months passed, the case was unresolved.

About every ten years after the incident, another detective would pick up the case file and start over on the case. New people would be interviewed, but nothing solid was ever developed. Because of the severity of the crime, every longtime resident in Great Falls was familiar with it.

Now fast forward to adulthood. Allen Reavley would often talk about it with his local acquaintances because of his association with the grocery store at the time. As an adult, Allen ran the local food bank for his employment.

He also offered weekend retreats through the Catholic Church. The retreats offered the opportunity to seek forgiveness, and as the lecturer, Allen would sometimes say that he had done terrible things when he was young and would elude to the old double homicide. Allen liked the attention he got from the association with the old crime. His knowledge of this crime appeared to make him feel important.

A few of the attendees at the retreats were suspicious of his involvement, and word eventually got to local law enforcement. A few witnesses out of the seventeen present actually thought they had heard Allen say he had committed the crime. Later the police called Allen in for an interview about his knowledge of the case. Allen often bragged about his alleged involvement, and now it was serious. Two detectives were looking him in the eye; he had to either confess all or admit it was a hoax. Allen was in a pickle. He struggled in the interview and was evasive on some important questions.

Allen made the statement, "I don't have any contact with any of those people, but it always has been my greatest fear that it's going to be solved one day; it's going to be somebody that I know." But in the arrest affidavit, the statement was changed to "It's always been his greatest fear that it's going to be solved." Purposely misleading?

At this time the detectives uncovered another surprise. As head of the local food bank, Allen was taking funds for his own use. This was to the tune of about $100,000. Allen was indicted for theft, later pled guilty, and was sentenced for that crime. A short time later, Allen Reavley was arrested for the deaths of James and Lois Arrotta, the store manager and his wife from 1964. This was big news in the community.

The prosecution sought to prove that Allen was angry for being fired and needed money. The prosecutor thought he had knowledge of the store safes, and it was also thought that Allen's girlfriend was pregnant, which was why Allen needed money. The state sought to prove their case by stating that after Allen returned home on the evening of the incident, he crawled out a basement window, went to the manager's house, and took the husband and wife hostage at knifepoint. He forced them to drive him to the store and retrieve the money from the safe. Something went bad at the store, and Allen killed both people and left with only money from the till. There was a blood trail leading from the back room of the store to the till up front and some blood on the till and safe as well. The prosecution theory was that Allen killed the wife, injured the husband, and forced him to the front of the store while he was bleeding, thereby leaving a blood trail. Both victims were tied up at some point, as the ropes were still on the victims when discovered.

Back in 1964, nineteen days prior to the incident, Allen did take some money from the store and was caught. The store manager came to Allen's house after the theft and talked with him and his father about it. After the visit, Allen was so upset and withdrawn that his father took him in for a psychological evaluation at a local state hospital.

On the night of the homicides, Allen's parents insisted that he was home early that night and was at home when the murders occurred. Allen, at that age, was slight of build, and it did not seem reasonable to me that he alone could convince two parents to leave seven children behind and force them to go to the store, let alone be successful at murdering them. Despite this unlikely scenario, the police now believed that he could have crawled out one of the basement windows of his house without his parent's knowledge and committed the crime. The fact that he took a polygraph in 1964 and was cleared by the first detectives in the case was not mentioned. There was also no mention of an FBI report that the bloody fingerprints on the front counter did not belong to Allen and were never matched to anyone.

One of Allen's old high school friends was a local attorney. His name was Channing Hartelius, and he was a very successful civil trial attorney but had never tried a murder case in his life. To his credit, he had a large modern office in the center of town with several attorneys, secretaries, and investigators. Despite this, he needed help with all the reports and the investigation. Channing offered his help to Allen for a small retainer, since he did not believe the case would go very far. But Channing followed up by hiring a recently retired detective from the local police department to assist him.

As the months passed, the attorney got Allen out of jail, but other than that, the case was not going anywhere in Allen's favor. There were forensic questions concerning some blood typing that was done in 1964 and the fingerprinting. Channing called the Montana state crime lab and asked if they knew anyone who could help him. One scientist knew that I had just retired in 2002 and gave him my name.

Channing called, and soon after, I was headed for Montana. I spent a week with the current investigator and the attorney talking about the case, reviewing a few recently generated reports, and setting up meetings with the staff. I made several suggestions regarding our investigation of this incident and returned home. In my absence, nothing was accomplished. I returned a few weeks later, went to the police department, and asked for some photos that reports suggested existed. To my surprise, they already knew what I wanted. It was clear to me that our investigator was still associated with the PD and was discussing our case with them. I was very unhappy.

I returned to the law office, walked in, closed the door, and asked Channing a very simple question. I said, "Who is in charge of this investigation? We have some serious problems."

Channing paused a moment, looked up at me, and said, "You are." That was the last day for that investigator.

At this point it was time to finish looking at all the discovery. All three thousand pages of it. I went back to the Charlie Russell Manor where I was staying and started opening boxes of documents. It took a week. I stopped for a few short outings, meals, and a few conferences with the attorney.

I do not recall how many people were interviewed in total, but it was too many to recall. I focused my time on people named in the first year. Usually in a small town, the police have a good idea of possible suspects based simply on previous contact. One had to also wonder if this manner of robbery was common during this time period. To my surprise, an Internet search showed that it was common. Usually the target was a bank. The robbers would go to a town, determine who managed the bank, and follow the person until they knew where the person lived. They would do what we call today a *home invasion*. They would hold the family hostage, take the bank manager to his bank, and force him to open the safe and hand over the money. These episodes usually ended with everyone alive.

At the grocery store, a new safe had recently been installed with a time lock that could be programed and could not be unlocked during the night. Surely there would have been some publicity about this type of safe to deter robbers, and who would not know? If a person did not know, perhaps this was a clue that would be useful.

Prison would be one of those places where access to the media might be difficult.

Looking back to the fall of 1963, a similar robbery had occurred at a business in Havre, Montana, a small town not far from Great Falls. Three men went to the home of the business manager;

one held the family hostage while two went with the manager to the business. They got $2,000 in cash initially, but two of the three were later arrested and convicted of the crime. How did they know the facts needed to pull off a similar robbery of a grocery store in Great Falls? What were the time lines? Where there any associates to provide information to the robbers? Information would reveal that a dairy-product delivery man knew the store and the owner but also was an acquaintance of the Havre suspects. How does this relate?

I needed an investigator. I was not trained to hunt down people although I knew what was involved. I contacted the best investigator that I knew, my childhood friend Steve. Steve had just retired as a detective for a sheriff's office in California.

Steve and I were inseparable as children. He spent most of our summers sleeping at my house, riding bicycles, and getting into mischief. Steve was an obsessive compulsive all his life, but he had the gift of gab. We must both be like that, because when we teamed up, it was so much fun. I knew that if I could get him on the case, he would become obsessed and even dream about it. Figure 48 demonstrates the extent of discovery in this case. Each binder contained hundreds of pages of information.

When Steve and I were in college, we sometimes hung out on weekends and would play tricks on people. He and I would pick someone out, walk up to the person, and see how much information we could get from them. Like with most young men, our targets were usually girls.

Reavley Case files

Figure 48

Role-playing for the ladies, we would claim that I was a med student and he was a law student. He was so good that I would give him a time limit to get a name and phone number.

It usually took less than five minutes. I can still remember his favorite line for the gals: "Could I call you in the morning about breakfast, or would you just prefer a nudge?"

Steve went into law enforcement and became a detective. His gift helped him get inside the head of suspects and witnesses alike, making him highly successful in his trade. He had convictions of several serial killers to his credit and could role-play with a con like he was their best friend. The last time we worked a case together, my wife, Steve and I were in a small town called Portola in northern California. We had finished lunch and were talking about our youth and the fun we had at times. A young man probably in his early twenties came walking up the street, and I told my wife that Steve could get all his vitals in less than five minutes. We let her hold the watch, and we both approached the young man.

First, we asked for directions. Then we asked did he live around here.

"Beautiful town," he replied.

"Bet it was fun living here," Steve said.

"Yeah."

"Did you live in town or out in the country?"

"In Town"

Then Steve said, "We are trying to find this person, said he lived around here. Ever hear of him?"

"No."

"Think your parents would know? They live around here?"

"Yeah, over on Fifth Street, small yellow house."

"OK. This guy was about your age; you must be about twenty-two years old, I would guess."

"Twenty-three."

"That's right; I think I saw you driving a Volkswagen earlier."

"No, I have a 1995 Chevy Camaro."

"Red?"

"No, tan."

"OK. Is it all right if we talk to your parents?"

"Sure."

"I presume you live there, too."

"Yup."

"What was that address?"

And we got it in three minutes flat in a stranger-tostranger contact on a street corner.

Steve arrived later that week from California, and I gave him the boxes of documents to review. Steve holed up in his room at the Charlie Russell Manor, to read the discovery. Like me, he just came out once in a while during the day for fresh air.

On one of those occasions, I asked him how it was going? He shrugged and said, "It is long and tedious, but the Macek investigation in the early seventies looked really promising. I wonder why they gave up on it?" Macek was an investigator who had researched the crime in the 1970s and initially identified potential suspects. This was something that needed further research.

I did not give him my thoughts on the case until he had a chance to formulate his own ideas on what we might be able to accomplish. But I was amazed; we both narrowed in on the same witnesses needing follow-up questioning.

"Macek was on it," Steve said.

"I agree," was my response. "Why would the DA look past this stuff in favor of Reavley? It does not make sense. We got the discovery from them, so they had to know."

The next day Steve and I visited the original Reavley home. In 2002, the Reavleys were no longer living in the same house as when the incident occurred, so we got permission from the current owners to see the house. The state contended that Allen could have left the house without his parents' permission or knowledge by just crawling out a basement window just above ground level. Steve and I went down to the basement of the house. The windows were about eighteen inches high by about three feet long. One would have to stand on a chair just to reach them.

What was important was that these windows were hinged at the bottom and only opened part way at the top. It would have been impossible for anyone to climb out one of these windows.

We discussed that point and believed that the police detectives probably made up this scenario without going to the house to look at the windows to determine if it was possible. If Allen had gone out a door, the parents would have heard him leave. I told Steve that the prosecution further speculated that this seventeen-year-old boy went to the Arrottas' house, caused both parents to leave their seven children behind, and have them take him to the grocery store after midnight.

While at the store, he killed both people without incurring a single injury to himself or a single bloodstain on his clothing.

I said, "This might have happened with a group of thugs, but not a seventeen-year-old boy with no prior assault history.

He knew the family, and it takes a hard-core thug to do something like this."

He agreed. Jim Arrotta was a well-built man; one could only speculate on how this might have gone down if Reavley was indeed the perpetrator and confronted the adults of the family late at night with a house full of children. He doubted that anyone would have left the Arrotta house except Allen in handcuffs. But what was more important was that the prosecution had not figured this out; they had to know.

From there, we did a viewing of all the evidence at the police department. It was important to know exactly what the prosecution had available and to actually see it. After forty years, it is not uncommon for some of it to come up missing. We needed to know exactly what was available. Stuff tends to disappear over the years, even in police custody. When we were in the company of the police detectives, Steve and I looked at each other; there was no doubt that we were the enemy in the eyes of the police department. Hard to believe after twenty-five years in law enforcement.

As I said before, the knife in evidence was important, in that the tip was bent, as seen in the earlier photograph of Figure 45. I had seen this in other cases; when the blade was thrust with considerable force and struck a solid surface such as a bone or floor, the tip would bend. In this case, that surface was probably the concrete floor in the rear of the store under the Mrs. Arrotta's torso. In these stabbing situations, it is also possible for the hand holding the knife to slip down the handle and contact the blade if the knife suddenly and unexpectedly stops. Depending on how the knife was held, the hand often was cut on the distal surface of the little finger or the distal fleshy part of the palm.

With this information, we traveled to the store to look at the floor. The knife had reportedly originated in the store and was used to cut vegetables. Cutting vegetables was not something that would bend the tip of the knife. Not much had changed in the backroom of the store; I had photos of the bodies and could locate the areas where they were lying when discovered. I got down on my hands and knees for a better view, and there I found several dents in the concrete that were consistent in size with the point of the knife. After forty years, they were still there in a location that would have been under Lois Arrotta. She had been found face down, and photos of her revealed incised wounds on her back that corresponded with wounds on her front chest. The injuries on the chest side were much shorter in length than the corresponding wounds on the back. The evidence supported the knife had completely penetrated her body. The person who did this, to this day, should have a scar on his or her distal palm or little finger of his or her strong hand, depending on how he or she held the knife.

I searched other areas of the floor and did not find similar depressions.

With this knowledge, I went back to the scene photographs and specifically the blood trail on the floor from the back room to the cash register. The prosecution believed that this was Jim Arrotta's blood, that he had been beaten and forced to walk up front to the cash register. This was probably not the blood trail of Arrotta but a blood trail from the assailant who had the knife in hand when the wife was stabbed. A sample of this blood had been collected but was consumed in the initial testing. The old report from 1964 had listed the blood type as type AB, the same type as the wife. It did not make sense that this was her blood leading up to the cash register. If she was bleeding enough to leave a blood trail, there should have been blood on the front of her clothing as well. There was not. James Arrotta was type A, if you believed the blood typing in the report. Why was the prosecution speculating that this was his blood?

The blood typing had been done in a clinical lab under the supervision of the local pathologist. I worked ten years in a clinical lab and was baffled how these people could type a dried bloodstain. We were not trained in these techniques; in the clinical laboratory, we worked with liquid blood. I had the name of the pathologist on a report; now forty years later, I had to find him and started making phone calls to local medical labs. A few days later, I found him alive and well in the Puget Sound area in Washington state. I called him, explained who I was, and refreshed his memory on the case. He remembered the case quite clearly. I asked how he had done the typing of the bloodstain, and he remembered how they did it but not the name of the technique. I asked if it was called the Lattes Crust technique, and he laughed and said yes that was it. I said thank you and told him I appreciated his help.

This was significant as another problem solved; this technique was a reverse type for blood. If someone like the wife who was type AB was typed, a negative result was obtained. In other words, there would be no difference between getting a proper result for an AB individual and the test not working at all. I am sure they knew the blood type of the wife from her liquid blood and just surmised that the lack of success in typing the dried blood was because she was type AB. This was not an uncommon mistake for someone not experienced in dried-blood typing. In our profession, the absence of a result should mean nothing was established or the type was AB. About 5 percent of the population was type AB. The door was open that the blood trail from the back room to the cash register could be from anyone. Too bad we had nothing left to test.

Apparently there had been a rush to get the store open during the crime scene search back in 1964. The scene exam had to be curtailed when that time arrived.

The chart below in Figure 49 was used in trial and outlines what was not collected from the scene.

Not Collected at Initial Investigation

- Wallet
- Candy bar wrapper next to Lois
- Tootsie Roll box not collected
- Bloody print at counter till
- Pop bottle at front counter
- Shoeprints
- Size 8 gloves
- Several bloodstains
- Assorted papers and coins at front counter

Figure 49

After viewing the evidence and looking over the few photographs available, I developed a theory of how this incident went down and went so badly.

Steve and I were excited about the information we had developed and our theory of the crime. We both had just retired from a career in law enforcement, and I had worked closely with DAs for twenty-five years. We needed to talk to the DA and explain to him our new information and that he was prosecuting the wrong person.

Channing said he doubted the DA would listen, but I was sure he would. With Channing's permission, Steve and I went to see the DA. We explained what we had or surmised. He listened, stood up, and pointed to the door. He said he had his man and to get out of his office. I was speechless. When we got back to his office, Channing was not surprised; he knew this fellow from prior exchanges. After that encounter, I set up a war room near Channing's office, and it was on; we were going to trial.

As Steve and I traveled back and forth between our homes and Great Falls, I would set up staff meetings in the war room with the group. We would go over our most recent findings, and I would make new assignments. We would be prepared.

Steve started contacting witnesses listed for the prosecution, as is customary in a criminal investigation. After he talked to the first witness, the ex-girlfriend who was believed to possibly be pregnant from Allen so many years ago, he got a threatening phone call from the head of the PD criminal division. The caller stated that if Steve contacted any more witnesses, he would personally arrest him for tampering with a witness. "And the arrest would not go smoothly," was the follow-up statement. Afterward he called all the witnesses and told them not to talk to us.

Normally the defense is entitled to talk to material witnesses in preparation for a trial—that is, everywhere but Great Falls, Montana.

Looking back on the boxes of reports, we saw that there was mention of two other robberies in 1963 with a similar MO. One was in Glendive and the other in Havre. In the Havre case, I am only going to use only first names of people in the reports. Gary and Larry were believed to have committed the crimes, but no arrests happened. In Glendive, a person by the name of Wilton reportedly was arrested, but Gary and Larry were listed as suspects. In both cases, the victims were tied up and left alive.

In the mid-1970s, John Macek had done some interviews with a guy named Wilber. Wilber had been released from prison a few days prior to the Arrotta homicides and lived in Havre. Wilber told Macek that just after he got out of prison, he was approached by a person named Gary about doing a robbery of a store in a neighboring town. Gary had a couple other fellows with him whom Wilber did not like, so Wilber declined to go. A few days after this meeting, Wilber left the state with a friend. While out of state, he got a phone call from his mother, and she mentioned the homicides in Great Falls. Wilber knew who did it and was not going to return to Montana. The last report from Macek stated that he was investigating four people who may have been involved in the Arrotta homicides. For some reason, the investigation ended, and we did not possess any further documents related to the investigation by Macek.

As stated earlier, on the night of the crime, one of the Arrotta children had remembered seeing a shadowy figure in the hallway outside her bedroom but could not remember any other details. This fact was reflected in reports generated a couple times over the years. A detective visited her in pretrial preparation and showed her a photo of Reavley in a football uniform.

She obviously knew who Reavley was, and the prosecution implied that they needed her to ID him as the person outside the door that night. Normally the police would provide a photo lineup with photos of several people on one page. The witness would be asked to pick out the one she thought committed the crime. The process for photo lineups are common throughout all fifty states—well, maybe forty-nine. Might I add that no incriminating statements are supposed to be made in a photo lineup? At trial, she had an epiphany and identified Reavley as the one she had seen in the doorway forty years prior. A sudden turn of events considering her prior statements. This was a surprise to all of us and possibly not good for Allen Reavley. The testimony was challenged, and the judge ultimately did not find it credible.

Since one of the reports in the box was a similar crime committed not far from Great Falls prior to this case, we wondered if there was additional information on the crime in the possession of the police department.

As usual, they would not cooperate with us. So Steve contacted the sheriff's office in the town where the crime occurred. Alerted by the PD, the sheriff knew we were going to call and would not give us additional information, either. He said as he ended the conversation, "We are not going to give you anything."

Again, we were the enemy. All cases can be difficult. The presumption of innocence is a constitutional right, but at this stage of the game, usually the facts dictate the direction of the case. Defending a person accused of a crime can be difficult at times as few are actually innocent of all association with an incident. The police usually get it right most of the time. When you realize you actually have an innocent person as a client, anyone who says defending an innocent person is easy is sadly mistaken. The pressure is on your shoulders; failure is not an option. The lack of cooperation with witnesses and the new charges against Reavley only made his defense even more difficult. The pressure is normally on the lawyers, but since I was leading the investigation, I was feeling it, too, despite the need to remain impartial. But we still had one ace up our sleeve.

Steve was on the trail of one of the guys we targeted in the reports. Both of us had seen the Macek data and felt it deserved more attention. This data was what had caught our eye several weeks before when we were first going through the discovery. The people of interest had committed similar crimes near the time of this incident.

We had no idea where he was or if he was still alive, but he had to find Wilber. With the names from the old report, Steve went back to the old court records from all nearby counties close to Great Falls. From names generated and subsequent interviews, Steve discovered that Wilber was somewhere in the Boise, Idaho, area and was gay. That is all he had when he drove to Boise. There he started frequenting the gay bars and asking around for Wilber. Steve could strike up a conversation with anyone. In the next several days, he ran across someone who knew Wilber and found out that he lived in a small town nearby. He also learned that Wilber drove a blue Volkswagen with a primer painted front fender.

Steve drove to the small town and just started asking questions and driving up and down the streets and alleys of the town. Hours passed without success. He eventually stopped at a service station and repair shop owned by a man of Russian descent. When Steve talked to him about Wilber, he was surprised when the owner called out all of his employees and had them stand at attention in a single line while he addressed them. One person permitted to speak thought that Wilber lived in a particular part of town but did not know the exact street or house number. Steve thanked them, and they were commanded back to work.

He continued to drive the streets and alleys for hours with no success until he came to a yard with a tall fence around the backyard. In order to see over the fence, Steve had to stand on the hood of his rental car; from there, he spotted the Volkswagen with the primer colored front fender. He went around to the front door of the house and knocked. Wilber answered the door. Steve told him why he was there, and Wilber invited him in to talk. Wilber repeated what he had told Macek so many years before. He knew it was Gary and his friends who had done the homicides; it was just too improbable that it was not. We called Wilber as a witness, and he testified to what he had told Steve.

We still did not know why the prosecution was avoiding the Macek reports, so this was the next step in pretrial preparation. The prosecution had dismissed Gary as a suspect. It was reported to us that Gary was working in a mine at the time of the homicides, so he had an alibi. The prosecution was probably hanging their hat on this fact, or was it a fact? This potential conflict of information had to be resolved, and Steve went to the mine, which was still in operation, and talked to the supervisor. He was able to access their time cards from 1964, which was a big surprise to all of us. Here the truth was found; Gary had been laid off or fired, not sure which, three days prior to the homicides. The prosecution information was wrong.

In addition, in 1972, fingerprints from the scene were sent to the FBI for comparison to Gary, but no match was affected to him or Reavley. Maybe Gary was not alone. In previous crimes where Gary had reportedly been suspected or involved, the victims survived. Did he have some new accomplices?

A follow-up investigation by Steve revealed that two of the three individuals believed by us to have committed this crime were dead. The other was living, and his location was known. There may have been others, but no leads were developed. When Steve returned to Great Falls. We held another meeting. It was decided that the DA would not respond to the new information, so trial would proceed as normal. We had time to prepare the trial graphics and witness list.

In reviewing the evidence and the reports, Steve and I had put together our own theory of what happened to the Arrottas that night. At least three people, possibly more, went to the Arrotta house and parked a car in the alley near the home. One stayed with the children while the others took the parents to the store.

The persuasive force was probably potential harm to the children; no weapon other than the store knife was recovered. A revolver was used in a prior robbery, but if they were packing guns, one would think use of a firearm to terminate lives would have been less personal than what really happened.

But some might argue there was the problem of sound with a handgun discharge.

They arrived at the store around midnight and were seen entering by a passing motorist. The motorist said he saw six people enter the store. Inside the store, as mentioned before, they could not access the safe, due to the new timer. Both the husband and wife were bound with white cord from the store, again a lack of preparation for this moment. Both victims were struck over the head with a glass bottle while on the floor, perhaps to encourage cooperation.

The murder weapon originated from inside the store, again a lack of preparation for murder, as they probably did not know it existed until the suspects were inside. It was supposed to go like the previous crimes: Get the money, tie them up, and leave.

As mentioned earlier, the sudden stop of the knife after passing through the body and striking the concrete floor could have caused the hand to slip down the blade and cut the hand of the assailant. The assailant's blood was probably what was seen next to both bodies in vertical droplets and in a trail leading to the front of the store. The old term was *low-velocity bloodspatter*; now we just call it *impact spatter*. I suspect that the assailant hesitated in the hallway (Figure 46) near the white cloth to care for the wound, but other possibilities may have also existed.

It could have been possible that James Arrotta was stabbed, creating spatter on the back of his shirt from the stabbing motion of a bleeding hand. In a multiple-stabbing situation, the clothing will soak up most of the blood, and spatter created will be of two types. One will be cast-off spatter from blood on the knife, usually traveling behind the person swinging the knife. The other is from blood pooled on the surface of skin or clothing from a prior stabbing. The hand holding the knife contacts the pooled blood and creates bloodspatter. This spatter is usually radial around the impact site. What I saw on the back of James Arrotta's clothing was scattered bloodspatter and could have been from a cut hand. It was less likely to be impact spatter from contact with the bloodstained shirt. Based on the path of the knife through the chest and the subsequent limited pooling of the blood on the back of his shirt, James Arrotta was on the floor when stabbed. It is interesting that he was stabbed from

two different positions of the assailant. This was determined from the examination of each incised wound. One could see the sharp-edge side and the backbone of a single edge knife.

The possibilities were that two persons committed the stabbing or the assailant changed position during the assault. Unlike a gunshot to the head, death was not instant from an assault like this, as one had to bleed out internally or externally before death.

This took a few minutes for the brain to lose enough oxygen for the person to become unconscious. In the meantime, the victim might struggle, and movement of the victim and or assailant should be expected. The absence of sudden death may have caused the assailant to attack again from a different angle. It is sad when you have been in this business so long that you even know this stuff.

We collect evidence at a scene in hopes that an explanation will present itself over time. My interpretation here was based on the best but not the only possibility. In science, we call upon Occam's razor, or the law of succinctness.

It is a law that generally states that when faced with competing hypotheses that are equal in other respects, the one that makes the fewest assumptions is more probable.

In my hypothetical, one must also consider that movement by the wounded assailant after the death of both individuals could have created the vertical droplets as the assailant moved from one body to the other. If the suspect was ever apprehended, he or she should have a significant scar on the palm surface of the strong hand. But in rare instances, some people hold a knife with the sharp blade away from them. In this situation, the cut would occur to the fingers. By the way, Allen Reavley did not have any scars on his hands.

In considering the opposing question of the vertical droplets of blood originating from either victim and in the absence of a good blood typing, the victims would have had to be upright and bleeding at the time the blood fell to the floor. In photographs of him on the floor, James Arrotta had blood coming from his nose, another remote possibility that cannot be eliminated. We will never know, since the stab wounds penetrated the lungs and he was facedown; the postmortem bleeding would exit the mouth and nose anyway by gravity.

During the scene search, the fingerprints in blood and the blood trail leading to the front of the store where the cash register was located were probably blood from the assailant, as mentioned before. All fingerprint comparisons to Allen Reavley were eliminations. Some bloodstains were typed and found to be type A, the same as James Arrotta. Lois Arrotta was type AB; she was excluded. Allen Reavley was type O and so was also excluded. Some bloodstains located in other areas of the store were tested and found to be type A. Forty-two percent of the population is type A.

Another item of evidence was a pair of elk-hide gloves found in an alley near the store. Any leather item was almost impossible to ABO blood type by the method used in Dr. Pfaff's lab in 1964. The absence of getting a result on the test would again correspond with a result of type AB. Therefore, the final result could be that the test did not work.

An examination of the glove interior produced a few hair fragments. These were sent to a mitochondrial-DNA lab back east in 2001. The results showed a mixture, suggesting at least two or more people had worn the gloves. None of these results implicated Reavley.

Some other dried blood remained in evidence and was sent to the Montana state crime lab for testing. The only request was to compare it to Allen Reavley. They were not sent standards from Jim or Lois Arrotta for elimination. The following items of evidence were not retested by DNA technology: The knife handle or blade, the leather gloves or the apron, pop bottle fragments, the key, the wallet, rope and blood scrapings from an envelope.

The crime lab did get results on some blood items, and as mentioned before, Reavley was eliminated.

What is especially odd is that the results were not entered into CODIS. CODIS is a national data bank that contains DNA results from thousands of individuals. I am still puzzled why this was not done, but it could still be done if the crime lab records remain. I must a confess that sometimes the prosecution will perform tests and the results never find their way over to the defense.

In theory, someone may have done the search, but I never saw the results. Figure 50 depicts the blood on the leather gloves. Testing in Montana by the state crime lab is free for the prosecution.

Why run selected items and not all of the items?

Was there a reason the prosecution would not want to know all of the results?

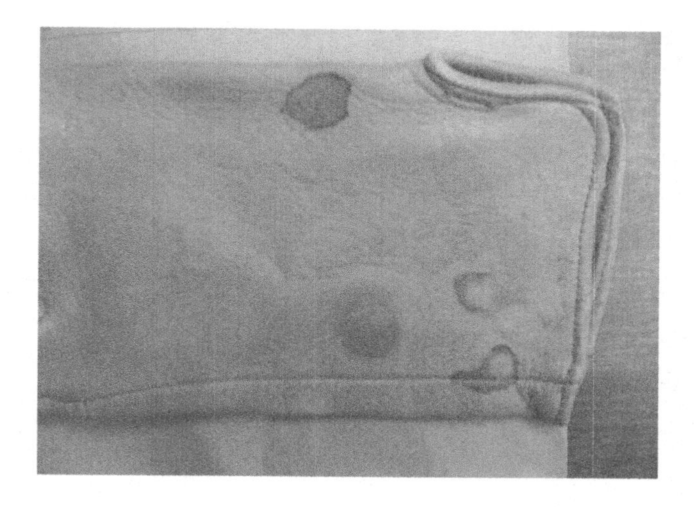

Figure 50

It could be that the prosecution did not know the details and the limitations of the blood testing done on the gloves and believed that Reavley probably wore the gloves during the commission of the crime anyway.

In a review of the scene photos, there was one that depicted the tied-up feet of James Arrotta, as seen in the next figure. Close examination of the photo showed the ends of cut cord in close proximity to each other. It was apparent that the cord was cut while James Arrotta was lying on the floor, Figure 51. One can only speculate why this cord and the cord binding Lois Arrotta's were cut, but sometimes inexperienced officers at a scene will go ahead and begin processing and thereby altering the scene before photographs are taken. I have seen it before.

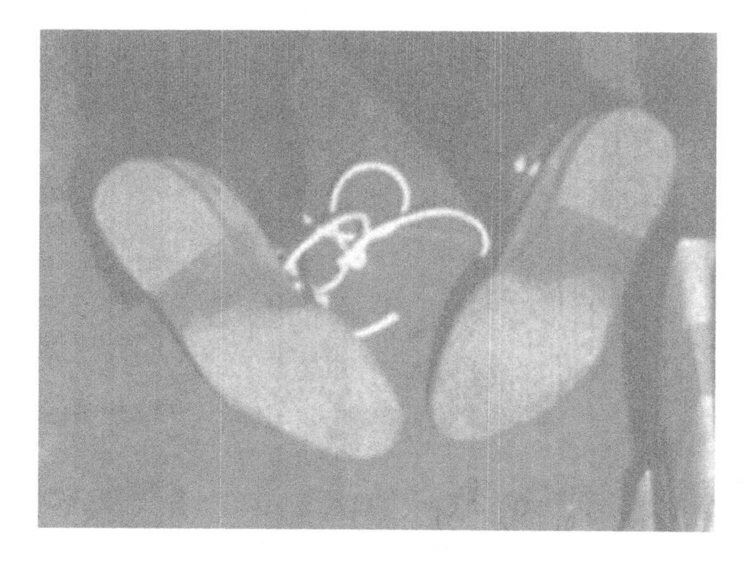

Figure 51

I digress again, but I remember one case when the police were processing a scene and it became apparent that they needed a search warrant. So they put all the evidence back where they found it until the warrant was executed. Then they picked up the evidence again. Somehow the first action did not make it into the police report of the scene processing. *Mum* was the word, or so they thought. To their surprise, across the street from the scene was a guy who was videotaping all the activities of the police, including putting the stuff back in the house. Besides getting egg on their face, the evidence was suppressed.

Back on the Reavley case, we knew that the DNA analysts from the crime lab were going to testify at the trial. But the normal protocol for the labs was that when a defense person calls to talk to you, you will call the DA and get his or her permission. We wanted to get around that, not to influence their testimony, which was probably not possible anyway, but to ask why some things were tested and not others. If possible, maybe they would give up a little bit about conversations with the DA. In most states, there is dual discovery with experts in a criminal trial. That includes notes, reports, and conversations if recorded.

I knew the people who did the work and knew they were good people. I invited them to stay where we were staying at the Charlie Russell Manor. They accepted, and we had a wonderful dinner and good conversation. They did not offer much in terms of any conversations with the DA, but we were able to tell them about our case. I suspected that the DNA people had probably heard from state investigators and read the newspapers as well. At least they knew there was another story, and we hoped only for impartial testimony from them. When the scientists took the stand, that is exactly what we got.

For experts in trial, we all know we are to provide opinions based on facts or the way we interpret those facts. One side or the other has retained us—for free, if it is the state, or for a fee, if it is the defense. The side that hired you wants you to do well and is your personal advocate; the other side hopes you will do poorly, thus enhancing their case. We take the stand and try to be relaxed and answer the questions. When it is time for cross-examination, we hope to exude confidence and try not to change our persona, but it seldom works.

Early in my career, I used to turn to the cross-examining attorney and cross my arms. My brain was saying, "Go ahead; take your best shot." Hopefully in time I got better at hiding my feelings, but I know my thought process did not change. We know that if the opposing side cannot attack the evidence, they will attack you personally if they can. We know it is coming, and it can be somewhat unnerving while waiting in the hall for the call to testify. But I also know that when I am attacked personally, there is nothing in the opinions about the evidence that they can attack. The DNA people knew that both sides would be friendly, as both had points to make.

When trial started, several of the Arrotta children were present for the trial. Since I sat outside in the hallway (as experts were excluded from the courtroom), I talked with some of them. No doubt this was a tragic event in any child's life, and no less for them. They were interested in getting justice for their family, and I could hardly blame them; it was just that in my mind the wrong guy was being accused.

Left to Right:
Darla Bergstad
Steve McCulloch
Channing Hartelius
Jeff Sutton
Tina Palagi
Julie Corbell
Jim Pex

Figure 52

Figure 52 depicts the defense team

My conversation with the family was light, and nothing regarding the upcoming testimony was discussed until everyone who had been called as a witness had testified. At that time Steve and I talked with a few of them about our findings in the case, which left them unsettled and not sure where to turn.

Steve sat in the courtroom for at least part of the case. The prosecutor had his detective seated near him in case he needed information, and occasionally the detective would turn around and make eye contact with Steve and try to stare him down. At first Steve just smiled back at him and laughed it off. He could tell that he was getting to the detective. Now the detective was really pissed.

The more Steve smiled, the more aggressive the detectives became in his body language. It was a good thing this was a trial before the judge and not a jury trial. These antics carried on for much of the trial with some threats taking place out in the hallway during recesses as well. It was clear that the two did not like each other. You think?

Wilber, the guy Steve found in Idaho eventually took the stand and told his story about the conversation he had had so many years ago and almost seemed relieved to finally get it out. He was a quiet sort, slight of build, and wore country-bumpkin style of clothes, if you want to call that a style. One of the first things he said when he took the stand was that he was a homosexual, and then he went on with answering other questions. I'm not sure how that related to this important testimony, but all in all, apparently, he was quite believable on the stand. The prosecutor was unable to budge him on his story, and he was finally excused with his statements untarnished.

I eventually took the stand and started with a PowerPoint presentation of the evidence and the scene as it was forty some years prior.

I went over the blood-typing information and the positions of the bodies and the knife. The knife, shown in Figure 53, was critical to my testimony. The dented tip in combination with the absence of a hilt was obvious to me when

I first saw the evidence that at least some of the blood on the floor was from the assailant. This, in combination with the few chips in the floor, was a combination that could not be ignored. There may have been other scenarios that fit this circumstance, but I had eliminated all that I could at the time.

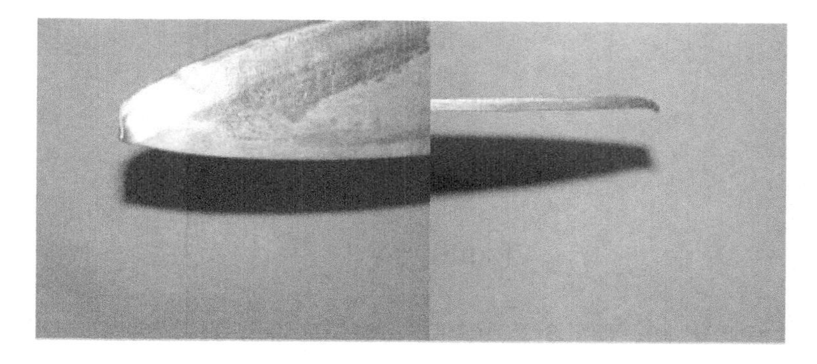

Autopsy Report

- "Three of the back wounds penetrate entirely through the chest cavity to the anterior surface extending to just beneath the skin. Two of these barely break the skin surface."

Figure 53

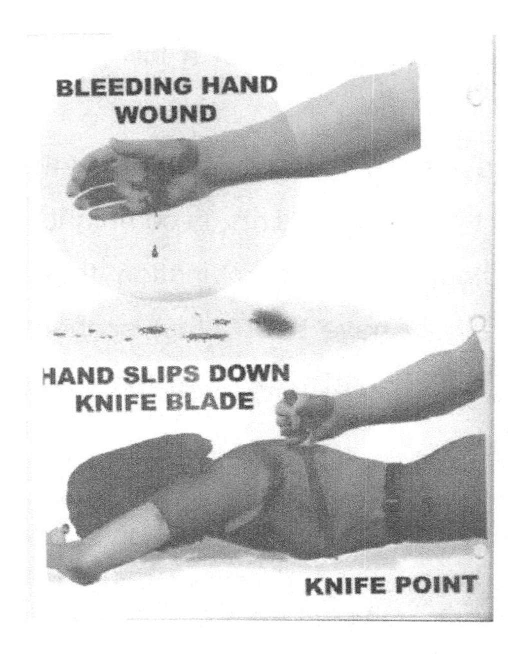

Figure 54

Figure 54 was a graphic used in trial to demonstrate how the tip was bent in contact with the concrete floor. It also shows where the hand may have been cut.

Below, Figure 55 and 56 are other slides from my presentation as an example of my testimony.

I had to do something to demonstrate how the bloodspatter on the floor was created, so I decided to do a live demonstration. The night before, I had drawn blood from one of the secretaries in Channing's office, and in court, I had white butcher paper placed on the floor.

Not Collected at Initial Investigation

- Wallet
- Candy bar wrapper next to Lois
- Tootsie Roll box not collected
- Bloody print at counter till
- Pop bottle at front counter
- Shoeprints
- Size 8 gloves
- Several bloodstains
- Assorted papers and coins at front counter

Figure 55

Reavley Evidence?

- No glass on Reavley's clothing
- No blood on Reavley's clothing
- No cuts on his hands
- No hair on Arrottas from Reavley
- No hair from Arrottas on Reavley
- Not Reavley's hair in gloves
- Not Reavley's sweat in gloves
- Not Reavley's blood type at scene

Figure 56

While the judge watched, I used a dropper and dripped blood on the butcher paper. The pattern of blood generated looked just like the large droplets of blood in the photographs.

It took several hours, but eventually I got through my testimony and was ready for cross-examination. I had covered all the main points, and there was nothing in terms of physical evidence that linked Reavley to the crime. On cross-examination, the prosecutor did not come after me with much determination. He concluded that I was simply wrong and that the absence of any evidence relating to Reavley did not mean he did not do it. I was eventually excused and was quite relieved to get out of the courtroom. In the hall, I met with the family members again. This time they were asking questions about what I had testified to and seemed more interested in the possibility that Reavley was not the one.

At one point in the trial, the prosecutor picked up the knife and held it in his hand during a demonstration of its use. When the defense pathologist took the stand, Channing asked him if the DNA from the prosecutor could be detected if the knife handle was tested. He stated that it could.

Then he followed with, "If the DNA of the prosecutor could be detected, why couldn't the DNA of the assailant also be detected?" The pathologist said there was no reason that it could not. As you have probably determined, the knife handle was never tested.

In finishing up, the last point the defense needed to make was the three people Wilber talked about could have committed this crime. The prosecution had counted on Gary, the prime suspect, was working in a mine on the day the homicides occurred. Therefore, they thought he was excluded. Steve went to the mine and discovered this was not true by reviewing the actual time cards, which was new information to the prosecutor and not well received.

The prosecution and defense finally rested, and the verdict was up to the judge, not a jury.

The judge wrote a document outlining the case and linked several of his significant decision points to my testimony. Reavley was acquitted by Judge McKitrick, in part, through my detailed forensic analysis of the crime scene and by the hard work of Steve in finding the star witness.

In any trial, the attorney and his or her experts should work together like violins in a symphony. Channing was a master at that organization; he was a pleasure to work with, and there were no surprises at trial.

> Steve and I believe that this case could still be solved. It is still possible that DNA testing of the blood could be compared to a national data base or to other family members of the known suspects. This alone could resolve this case. In addition, I would suspect that the one who did the stabbing would have a significant scar on his or her strong hand, most likely on the distal fatty portion of the palm. It is our hope that someday someone will take this case up again and resolve the issues through modern forensic science.

Chapter 21 Working with the Military

Criminal Defense

I had received several requests for expert witness consultation from the army and managed to negotiate the complicated federal system to set up payment. Not an easy task. Doing the cases was easier than negotiating the payment process. But I really liked these cases and the esprit de corps of the army courts-martial process. In most cases, we would start trial promptly at eight in the morning, and trial continued until noon. If anyone needed a potty break, the judge would stop the trial for that purpose and resume as soon as the person returned. One day the trial was moving behind schedule, and I saw the judge tell both sides to order in pizza as he was only going to take twenty minutes for lunch. In the military, the jury is called a *panel* and is composed of military personnel. Normally you will get a group of people who are conservative and knowledgeable about military conduct. They know firearms too, which is helpful at times.

After lunch, trial would resume, and we might go until the judge decided to stop. This might be at seven at night.

Then you went back to the office with the attorneys and began preparation for the next day.

That might take until midnight. The military system is special, in that all attorneys and the judge are often in the same set of offices. An attorney might be a prosecutor for a while and then be assigned a defense case. I met some who were very accomplished criminal defense attorneys. Then there were others who struggled with basic issues at hand. It was not uncommon for an attorney to ask me what he or she should do, especially in the latter group. A very strange role for a forensic scientist.

I had experience in hundreds of trials and worked with a large number of attorneys, so at times I could be helpful. In one instance, when I arrived at the army base, the attorneys rushed up to me and said they were so glad that I was finally there as they did not know how to proceed. So I would handhold all evening as we got organized, and the next day I would either sit at counsel table or directly behind it and pass notes on to the attorneys giving them some questions to ask after listening to a prosecution witness on direct exam. There were several occasions when I had received an expert's report and was asked to write the questions to be used in cross-examination. I would write several pages of questions and tell the attorney to pick and choose what questions he or she wanted to use. I also wrote the questions for the attorney to ask me when I was on the stand. At times we prevailed, so I must have been on target once in a while.

When I was on the witness stand, it was the responsibility of the attorney asking the questions to enter any gestures made by me into the court record, not just what I said. I am a gesturing person. I talk with my hands, and so while I was speaking, the attorney had to speak also in order to get all the gestures down for the record. It turned humorous when I asked one of my attorneys to remove his class-A jacket and lie on the floor so that I could demonstrate how the victim was shot. He had to get permission to remove his jacket, and then I had an attorney tracking all of my movements, down to putting one hand in front of the other, while I was also talking trying to explain the shooting. Then there was cross-examination, and we did it all over again.

It might go something like this: "Let the record show that Mr. Pex raised his right hand, pointing his index finger toward the attorney on the floor. The attorney on the floor, without his class A jacket, moved his right leg while Mr. Pex removed his left hand from his pocket, now making a right hand in the shape of a firearm. Immediately after raising his right hand, Mr. Pex said, 'Bang!'" I applaud these people for trying to keep up and still get their point across. It took a lot of extra effort. These were very serious proceedings, but it was difficult not to laugh at times.

I had the overwhelming desire to make a few esoteric gestures with my middle finger and then listen to how that went into the record, but I didn't.

On one trip, I had finished my testimony at a military base in Kansas and had to catch a flight early the next morning. I left around 3 a.m. the following morning in a small rental car, traveling to St. Louis. It was tornado season, and I headed east in the dark on the interstate. Based on the size of the hail and the reports on the radio, the weather outside was prime for a tornado. I could see lightning in the distance on both sides of I-70 and had no idea where I was at any given time. The sky was dark; lighting flashed as it rained or hailed as I traveled. I listened to the radio for any tornado information and realized that they announced the tornados by county. At any given moment, I had no idea of what county I was in! I was driving fast and was the only vehicle on the freeway. That in itself was a bad sign. Finally I could see the dawn ahead and was never so glad to see daylight. I reached my destination without further incident. Such were the adventures of a consulting expert witness.

Business continued to be brisk, and I was traveling all the time. The army was going to send me to Iraq on a couple of different occasions. But about the time we were ready to go, it was determined to be just too dangerous to make it happen, and plans were canceled. I did receive orders from the army to go to Fallujah to review a crime scene where a soldier went nuts and shot three of his roommates.

I traveled directly to Kuwait City late in the evening and went to the private airline that flew to Bagdad. They determined that my papers were incomplete, and they would not let me on the plane.

This was a predicament I was not prepared for, and I did not know where to go or even how to get from the airport. I was concerned, I might have to sleep in the airport. The military people I needed to talk to were not in until the morning.

Then some soldiers noticed me wondering and asked if they could help. I was very happy to see them and told them my story. They helped me get through customs and into a taxi, which took me to a local hotel. The hotel was a large white building in a secluded neighborhood where government contractors normally stayed when passing through the country, so this was good. My assigned room was quite large, and I had a huge bed. It was larger than a king size. I had to wonder how many people could sleep in this bed at a time. The following morning, I was on the cell phone and got in contact with the travel people and explained the problem. I had to have the name of a person who was waiting for me on the ground when I landed. For some reason, they did not know that and would get back to me in a couple days.

A couple days? What now? I went downstairs to the lobby for breakfast and noticed several other men hanging around alone. Like I said, contractors moved through there on their way to and from Afghanistan and Iraq.

I noticed one sitting by himself at a table, so I asked to join him. We struck up a conversation, and it turned out he was an interpreter for the army, a Coptic Christian from Egypt.

We had another fellow join us, and he was CIA; I think his name was John Smith, if you want to believe that. After breakfast, we hailed a taxi and went uptown to see the sights. Kuwait City was a bustling place with lots of construction and nice cars and was quite modern. I learned that one-third of the population was from somewhere else and came to Kuwait for work. The taxi drivers were no exception. They were a host of information about everything going on in any country or on any topic. I found that talking with these guys to be one of the best parts of the trip. I met drivers from India, Pakistan, Brazil, and Europe in my travels.

The interpreter wanted to go to the market, so we tagged along. It was an open-air market and very clean with quality vegetables and dry goods all about. Nothing was hidden, and all goods were out in the open with no security.

As we wandered, the spook and I passed the time hitting up the interpreter for local information about the sites and the people. At lunch, we stopped at an eatery in the market, and each of us ordered a salad. We got what we ordered, but to my surprise, we got a large salad and three forks, no other plates.

Oh, well, I was hungry and did not want to look too much like a tourist, so we jumped right in.

A few days later, I was still in Kuwait and now riding with a group of guys in a van. We were all quizzing each other on what had brought us to the city. At least they were older, and I did not look so out of place—that was, until we went by a pharmacy. Someone yelled to the driver, “Stop!’” He pulled over, and everyone ran for the pharmacy. I tagged along to see what the excitement was about. In Kuwait, you could buy Viagra without a prescription, and these guys were loading up. With pockets full, everyone was ready to see more of the city. I have to admit, the trip through the city was really interesting; I felt like a hobbit on an adventure.

As the army had said, it was a couple of days before I got my orders complete, and then I went back to the airport. Only one private airline would fly in and out of Bagdad, and it only traveled at night. Getting on the flight was routine, and the flight was OK except when we started to descend. They turned out all the lights in the airplane and advised us to hang on; we were going to descend rapidly. No kidding. We lost several thousand feet in minutes. I looked out the window at the yellow lights of the city and noticed bright flashes near the plane. Damn, were they shooting at us? I turned to the woman sitting next to me with wide eyes. She saw my look and seemed to know what I had noticed.

She said that there was an attack helicopter above and below us and that they were shooting off flares to confuse any missiles that might come our way.

That was comforting, or not. What had started out as an adventure was suddenly getting serious.

Once on the ground, the passengers were lined up in a military manner, and we were assigned our escorts to the camp where we would stay for the night. I was given a helmet and bulletproof vest and taken to a city of cheap aluminum travel trailers interspersed among huge pillars of concrete. Sort of a concrete hedge maze. The pillars of concrete looked like road dividers but were ten feet tall. This was my sleeping quarters within a maze, and I was the mouse. Once assigned my trailer, which had only a bed and a locker, I asked about the safety of this place. The soldier said they get attacked once in a while with missiles, but they are seldom successful. Basically, if your number was up, it was just bad luck; there was no other place to go.

The night was quiet. I know; I was up all night. When I finally had to pee, being the mouse became reality.

Figure 57

A photo that should instill fear in the enemy!

The Pex mouse was successful; despite a few dead-end turns, I finally found the restroom.

In the morning, I was directed to a huge hangar, which was the meal center. The hangar was full of people from all over the world. You could see uniforms of many different colors, hear foreign languages, and step into a chow line of your ethnic choosing.

I was not picky; I just wanted to make sure I could recognize it before I asked for a serving. I had seen these hangars on the news years ago when the United States was bombing Bagdad during Desert Storm using bunker-busting bombs. Now they were all renovated.

That morning I met with my travel guide, who was an army CID agent, an investigator type assigned to take me to Fallujah. These fellows were prosecution guys, and I could tell that he was not happy to babysit a defense expert to the compound.

At least he was not antagonistic and eventually lowered his shields, and we could talk about the case, politics, or sports. The military had spent a week at the scene and examined it from every different angle. Their reports were outstanding, and I told him that I was there to uphold the rights of the accused; I was not there to make up stuff for trial. If everything was done right, then so be it. He felt more comfortable with that statement.

We waited for several hours, and finally we were escorted to a helicopter. Here I was in civilian clothing, helmet, vest, and a suitcase. I must have been a sight for the locals. I was worried either that they would think I was very important to be here at my age or that the army was very desperate. I was not sure which one. I sat in the outside seat by the door, or should I say where the door should have been.

The door frame tucked slightly under the seat, and the only thing keeping me in was the seat belt. We took off in a hurry and climbed straight up for at least a thousand feet to get away from any snipers quickly and then started for Fallujah. As we passed over the desert, I could see the effects of the last war. There were huge bunkers blown up and sinkholes that had once contained Saddam's tanks. Now all that remained were spare parts scattered about the landscape. While in flight, a soldier was manning a machine gun next to me and another on the other side of the helicopter. How comforting.

I was recalculating in my head how much I was making, and it was not adding up at this moment. We landed in a field and immediately went inside a troop carrier with no windows and moved toward the base.

Space was really tight, especially with a soldier manning a machine gun in the middle of us. Once inside the fort, I could not believe where I was working. This was a compound with eight-foot-high brick walls and several secure buildings. It looked more like a prison than a base. Approximately forty soldiers stayed here at any given time. I asked if it was possible that someone could shoot at us over the wall, and the response was, "Sure." No going for walks in the evening.

A few days before, I had called my travel agent to confirm my return ticket. She had arranged for the extension of time, and all was set. I had first-class tickets for the return and was not going to pass them up because of army delays. If you don't have a travel agent, get one. She had saved my butt more than once. This adventure was no exception.

I explained to the CID guy that I would need to work all night, as I had to be on the next late-night flight out of Bagdad. He unlocked the crime-scene room and mentioned something about the waste of money, because they had done everything that could be done and it was all proper. In the time I had available, he said there was no way I could redo the scene to the quality that they had done. Perhaps, but time would tell.

In my work, I did not have to reinvent the wheel, only to see what was done and if it had been done correctly. I was to take lots of notes and photos, and I did not need a week. The scene was a large room with four bunks and assorted military gear. There were lots of bloodstains, bullets holes, and other related materials. Within one hour, I pointed out a miscalculation in a trajectory, and within two hours, I had found another. The agent agreed when I explained to him what I had seen, and he was getting more appreciative of my skills. The errors were not going to change the outcome of the trial, only improve on the work already done. I could have said nothing and let the lawyers ambush the prosecution witness, but I didn't. The opposing counsel would get my report anyway. Figure 58 is a graphic that demonstrated bullet trajectories within the room.

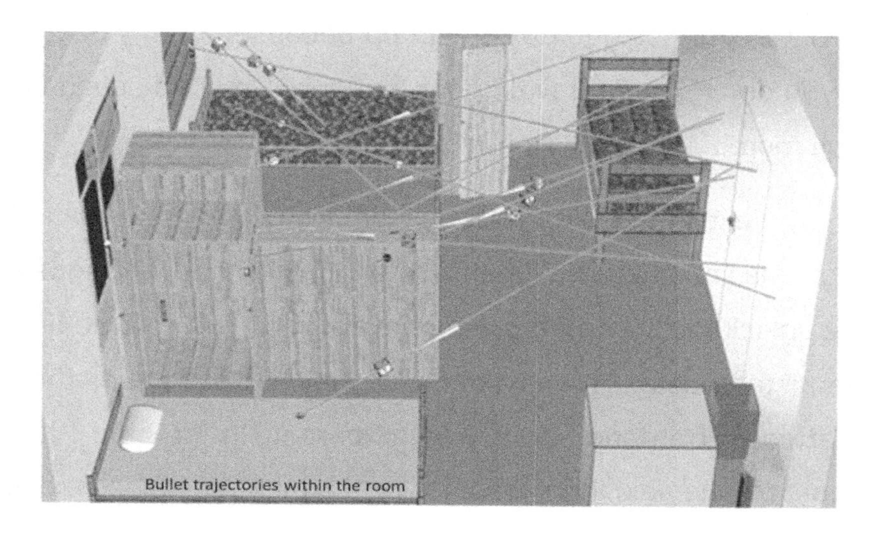

Figure 58

When I was done, I went to see the sergeant about sleeping quarters, and he just looked at me like that was not his problem. I went back to the CID agent, and he found a bed for me. I had no blankets, just a mattress and springs. I asked the sergeant for blankets, and he tossed me an old raincoat. It was cold that night, and all I had was the raincoat and the clothing in my suitcase. I opened the suitcase and started putting on all my clothes that I could get on. Then climbed onto the mattress, used my vest for a pillow, covered myself with the raincoat, and fumed. I was pissed. I did not deserve this kind of treatment from them; I was not a threat.

By next morning, we were out of there and back to Bagdad. I got the night flight back to Kuwait City and on to Frankfurt, Chicago, and Portland. It was good to be home after a week in the Middle East. There would be much more work to be done in this case.

A year later there was an attempt to get me to go to Afghanistan in the Bales case. Bales was a soldier who reportedly went into a village and started shooting everyone. By the time he returned to the base, he had killed seventeen people. For us, the village was not a safe place. In order to get the defense team in there, it would take major military force. As would be expected, the locals were not particularly happy with Americans after this incident. We learned that the prosecution team had not been to the scene, either, so we joined together for a single trip.

We wanted to use a special laser instrument that could quickly measure the entire scene with a few millimeters of accuracy. None of us on either side knew how to use it, so the company was going to supply their own people. The number of folks seeking to go to the scene was reaching one hundred. There were several potential windows for travel, but it just did not work out because of safety concerns within the villages. I was not disappointed. Bales pled guilty to all charges anyway.

There were many other military cases in which I consulted. In chapter 6, I touched on a portion of the Martinez case, in which we studied what happened to the handle when a hand grenade was thrown. Martinez was reported to be the first fragging case since the Vietnam war. *Fragging* is a circumstance when a senior ranking officer is killed by one of his troops, usually with a hand grenade. In this case, a claymore mine was placed in the window of an officer's room and detonated. Two ranking military officers were killed. I spent several months going back and forth to Fort Bragg on this case. Eventually the gate security would recognize me and quickly pass me through. After a lengthy trial, Martinez was eventually found to be not guilty of the offense. This was a fascinating case, and a book could be written on this case alone. Not all of the military cases had problems, and for the most part, they were pretty straightforward. The JAG officers and staff I worked with were above reproach.

They were some of the finest people whom I have met. When there were problems, there was a uniform effort to solve them instead of looking for blame. This was an attitude I liked.

The military had a crime lab near Atlanta, Georgia, a place that I frequented on occasion also. The staff was always helpful with very little antagonism toward me as a defense expert. Often, I would have more information about the case than they did, and we would discuss in general terms what was happening and if there were trial expectations. I would usually observe lab staff perform tests, offer suggestions, and provide the attorneys with information about the testing. In Martinez, there was a concern for the presence of trace elements from the claymore and hand-grenade blasts. No one really knew what to expect, so together we contacted the manufacturer to get details in their elemental composition. Later with the help of an explosive expert, we detonated both hand grenades and claymores with trace-element collection stations set up nearby. This was just another example of interesting problems I confronted with the military.

In another case from Iraq, a small platoon of soldiers was accused of executing local residents on an island in the Tigress River. I received photos and written reports to review, and the trial was to take place in Kentucky.

On the flight to Kentucky, I was sitting with the consulting pathologist, and we started comparing notes.

He had the autopsy report, and I had scene photos. It quickly became clear to both of us that the deceased in the photos was not the person who received the autopsy. The person in the autopsy had lost a leg due to an explosive device, and that was the cause of death. In the case photos, the deceased had both legs.

Again, you can't make this stuff up. When I received a new package of reports in the mail, I just never knew what surprises might be in store. If there was one common theme, it was: *take nothing for granted*.

Chapter 22 The Danny Schultz Case

In 2011, I got a call from an attorney living in a rural area of California requesting my help defending his client.

Danny was short, about five feet eight inches, and slim. He was a Vietnam veteran and came back from the war with a bunch of problems we now call PTSD. He moved to a small acreage just outside town, raised some chickens, and lived pretty much to himself for many years. The townspeople thought he was odd, and Danny was seldom seen mixing with other folks. I was told he had been a machine gunner in an attack helicopter and saw quite a bit of action. It was well known that life expectancy in that job was short, but he survived to return home a much different person from when he had left, not an uncommon occurrence.

A stranger came onto his property and asked for work. Danny did not have much money, but he told the guy if he split a pile of wood for him, he would pay him for the work. The guy split some of the wood and then demanded payment for all of the wood. Danny refused, and the stranger head-butted Danny, which knocked him out. The guy was six feet two inches and heavyset, and unknown to Danny, he had recently been released from prison for an assault on another individual. That individual ended up in the hospital. Danny was unaware that he was dealing with a dangerous felon.

When Danny regained consciousness, the guy was still on his property, so Danny ran into his trailer and grabbed a shotgun. Danny pointed the shotgun at the stranger and ordered the him off the property. What happened next was up for discussion. Danny shot the guy in the throat, and he died at the scene. Then Danny called the police. The police responded to the scene to find Danny holed up in his trailer; he had to be ordered to come out unarmed. He was afraid of the police. Danny occasionally had problems with alcohol, and this had resulted in brushes with the law. Here was a weird guy, known to use alcohol, who had a gun and was unwilling to come out of his residence. In this circumstance, the police had a right to be concerned and approached the scene with their weapons in hand.

From Danny's perspective, he was thinking, *What are all these guys doing out there with guns in hand when this is self-defense, and are they going to shoot me?* He eventually and reluctantly stepped out of his small trailer that he called his home and was taken to the hospital for his injuries and an interview. Danny clearly stated what happened, and the case was considered self-defense. Danny had a significant injury to his forehead to support his claims.

During the interview, Danny was consistent as to the distance between the victim and him when the shot was fired. He said about seven to ten feet was between them; the guy was approaching and would not stop. Danny feared for his life and fired.

To understand this case, you need to understand shotgun ballistics with pellets. Shotshells contain a large number of small pellets. When a shot is fired at close range, you just get a hole where all the pellets pass through. As the distance increases, the shot pattern expands, and individual pellets start making single holes in the target until the distance is sufficient that all the pellets make single holes in the target as depicted in Figure 59.

The pellet pattern diameter increases with distance

Figure 59

An interesting fact about pellets is that they line up behind one another in the shotshell, and at intermediate ranges, some may pass through the same hole as the pellet that stopped in front of it. To make matters more complex, when pellets strike solid objects such as tissue, they may strike the pellet in front of it and ricochet to the side; thus, the final pattern when observed in an x-ray is considered to be larger than the pattern would be if you fired at the same distance through paper. Medical examiners and firearms examiners know this from their training and experience.

The victim was autopsied a few days later. The pathologist reported that the entry was a round hole; when the skin was folded into its original position, the hole diameter was three-quarters of an inch. The spread of the pellets was just the ping-pong effect common to these shootings. When compared to test-fired patterns with the same ammunition and weapon, the distance was closer than Danny described. The report stated that the distance between the muzzle of the shotgun and the victim was five feet or less. That was the maximum distance for this weapon to just make a round entry hole without any stray pellets around it; it could have been closer.

The information was passed back to the agency, and the new inexperienced DA and the detectives decided that Danny had been untruthful and had him arrested for murder.

He was handcuffed and sent to the county jail. In a hearing, the prosecutor said that Danny sought out the trespasser and shot him intentionally in the throat knowing this would be a fatal wound. You got to wonder where they think of this stuff. *Intentionally* was the key word here, despite Danny being the only witness who stated several times it was self-defense. Then there was the goose egg on his forehead where he had been head butted by the assailant prior to the shooting. The absence of any robbery or property damage added to the DA's argument. Despite the injury to Danny, the prosecution argued there was not sufficient evidence of a struggle to warrant the use of deadly force. At this point the defense attorney used a phrase in the hearing that is not often heard in forensic science: *The absence of evidence is not evidence of absence*. This was a surprise.

I got a call from my friend and former prosecutor-turned defense attorney Jordan Funk. He believed his client, which was unusual for him. He was a pretty savvy guy and was not often fooled. I told him over the phone that in most self-defense shootings, the distance was going to be five to seven feet, from muzzle of shotgun to target. This was before I had seen any of the evidence. There are a couple factors here: The distance a client gives you is only relative based on his or her mental state at the time. The distance people usually quote is from their eyes to the eyes of the assailant. Proximity-testing numbers are from the end of the barrel to the target surface.

I have observed in several shootings cases that there is a point where a concerned person fearing for his or her life will shoot. When you consider the length of the outstretched arm or length of the firearm and the length of the arms of the assailant, the shot will be fired just before the assailant can get his or her hand on the weapon and put the other person's life in jeopardy. Let's add this up; the length of the arm of a large male is about three feet. The distance from eyes to the end of the muzzle is at least two feet.

A frightened person is going to shoot before the attacker can reach the muzzle of the weapon.

That distance is consistent to be five to seven feet from muzzle to eyes of the other person: six to eight feet from eyes to eyes. It might be a little farther if the assailant is coming at you quickly. In animals, we call it *fight-or-flight syndrome*; this is the point where a pheasant will flush when hunted. The pheasant can no longer outrun the predator and needs to hide or flee. I have researched this phenomenon extensively and have never seen a reference for this moment in time for humans. The distances I stated are based solely on my observations. I would love to know more about this decision-making process in humans. It is important.

My wife and I drove to California and met with Jordan at his home office, as is common in rural areas. Danny had been released on bail, so I got a chance to talk to him and view the crime scene. I took a lot of measurements and photographs for our own use. From this evidence, we generated several graphics as seen in Figure 60 that depicted the shooting scenario as described by Danny. The police were right; their facts of the shooting distance did not match up with the scene.

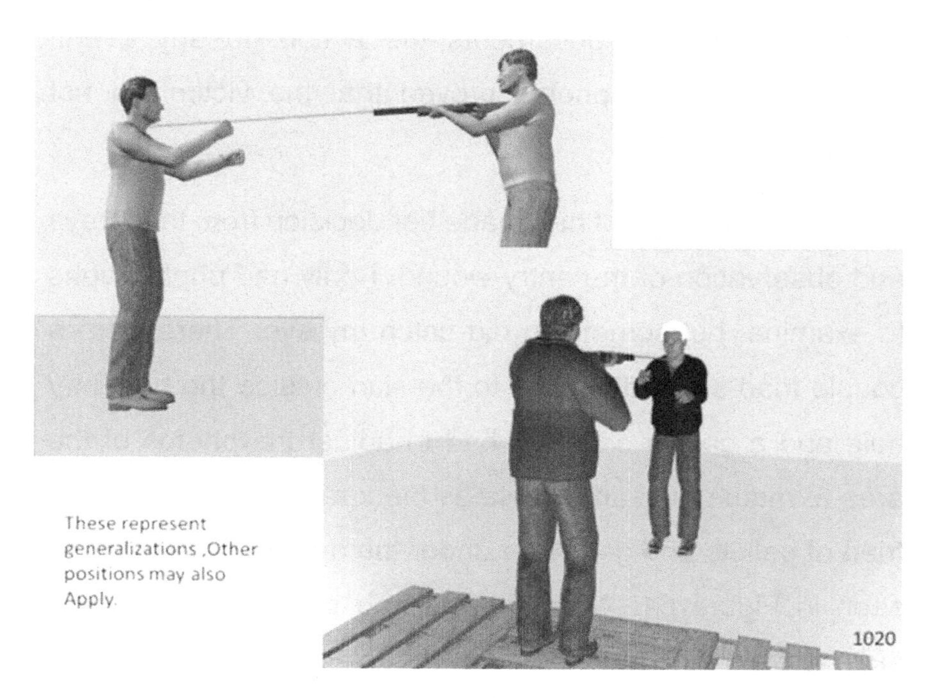

Figure 60

But Danny stuck to his guns, no pun intended, and was certain that his description of the incident was how it happened. I examined the clothing of both people and found nothing of consequence. Then I studied the police photographs taken at the scene with attention to the bloodstains on the deceased. It did not look like the victim had been dragged to his final location and appeared that he had fallen to one knee and then onto his back.

But you cannot always rely on photos for the final position, as it is common that EMTs will roll the body over to check for a heartbeat before the photos are taken by the police and no one documents the action. In any event, bloodspatter in the photos proved that the victim did not move.

The pathologist had made her decision from the x-rays and observation of the entry wound. I only had photographs to examine, but something did catch my eye. There were a couple triad-shaped injuries to the skin outside the big entry hole and a gap in pellets. I had to adjust the photos of the area to match the same scale as the x-ray. Oops! The same triad of pellets and gap were under the defects in the skin as seen in Figure 61. A few pellets had penetrated the neck outside the main entry hole. This meant that the distance was greater than described by the pathologist. In the photos, I could see where there were areas of dried blood that were not cleaned up well. This area was one of them and may have confused the pathologist.

I did my own proximity testing, compared the testing to my new information on the pattern diameter, and established the distance to be in line with what Danny had said, not what the crime lab suggested. We released our findings, but law enforcement was unimpressed. We would just duel it out in trial. In support, we hired a pathologist to review my findings, and the new pathologist agreed with me (Figure 61).

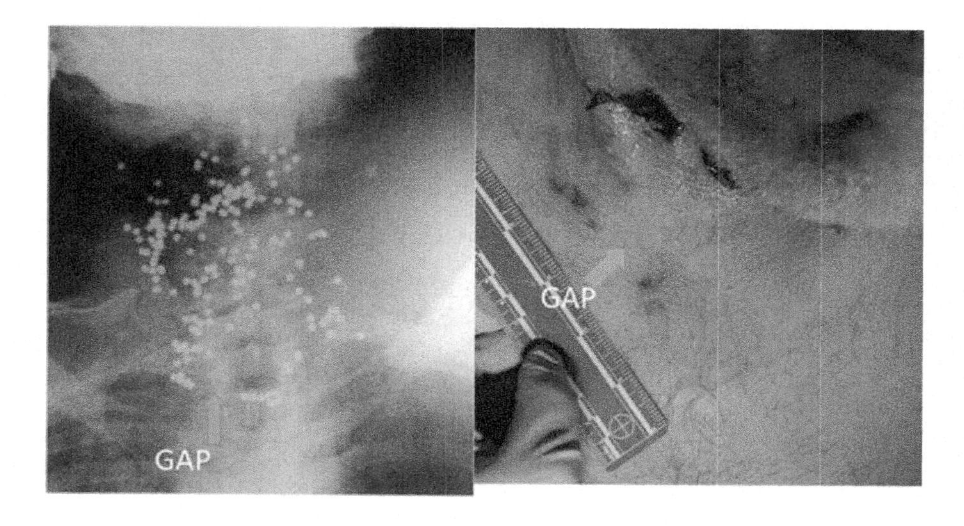

Figure 61

It is rare to discover through forensic science that the prosecution has got it wrong, but it happens. The burden on you as the expert is considerable when the client is facing twenty years in prison and you hold the keys to his freedom; everything is on you.

Jordan was worried. How did you get a jury in a small town where everyone trusted law enforcement and everyone knew that Danny was a basket case to be impartial? All Danny wanted was to go back to his little farm and his chickens. It was sad, but while he was in jail, the county hauled off his chickens and destroyed them. Eventually Danny was deemed not a flight risk and was released on bail prior to trial. He was devastated to find his chickens gone.

I traveled back to California a few months later, and our team met the evening before trial and discussed how to proceed. The next day a jury was picked, and we got on to business. Experts were allowed to remain in the gallery, and we got to hear all of the evidence presented by the state. Jordan, my wife, and I had lunch with Danny. It felt kind of weird to be at lunch with a murder suspect, and I voiced my concern to Jordan. He said, "No big deal," and we continued with lunch. I felt OK about it as long as the defense attorney was present. In my private practice, I have met with defendants on many occasions, often they were the only witness and a source of information. The evidence might not support their statements, but it is important to listen to them anyway.

Now it was our turn. I had to-scale charts of the farm, graphics on the position of the assailant and Danny, charts of the proximity testing, and additional photographs. Figure 62 was another software approach to describing the scene.

On the stand, we went through the testing, and I pointed out the discovery of an error in the autopsy findings by the state and had enlarged photographs to show it. Our pathologist took the stand after me and confirmed my findings. Our pathologist started out calm and professional, that is until you irritated him, and then watch out; he was no one to mess with. The young DA found this out firsthand.

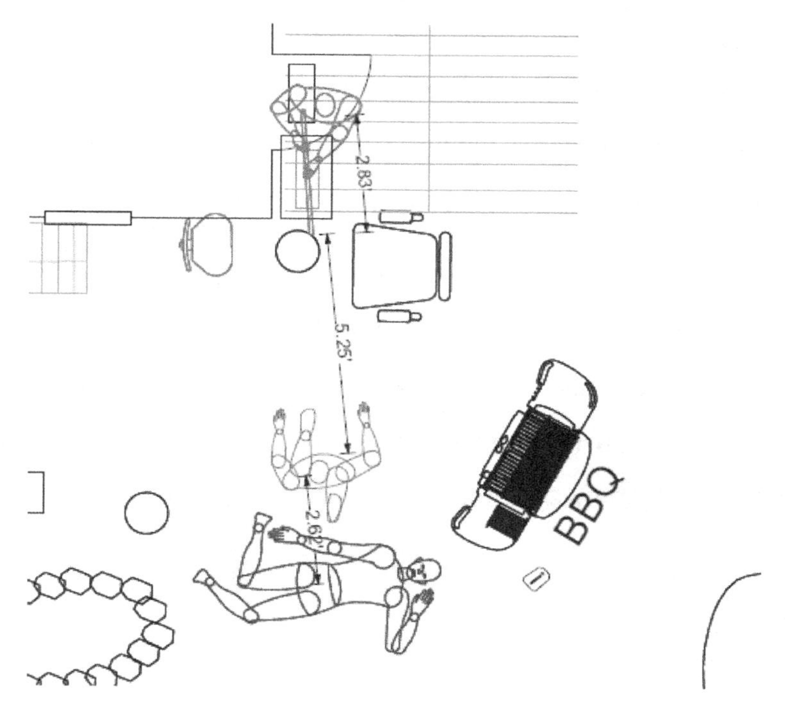

Figure 62

Closing arguments were heard, and the jury received the case. As I recall, they were out only a short time, and we all hung around hoping that the deliberations would be short. We were emotionally drained, and our hearts went out for Danny. Under the circumstances of the shooting and the evidence we presented, this had to have a happy ending. The court reconvened, and Danny was asked to stand for the verdict and it was read. Not Guilty! Wow, what an emotional moment for all of us. Jordan was ecstatic. Danny was reduced to tears, and you would never guess what happened next! The jurors came down out of the box, hugged him, and cried with him.

Danny came over, shook my hand, and thanked me for all that I had done for him. My wife and I were caught up in the emotional moment as well and could not hold back the tears. I had never experienced anything like this in my career.

People sometimes ask me how I can actually work for those defense attorneys and defend an obviously guilty person. You only need to see this happen one time, and everything you have ever done in your career is suddenly worth it. Danny walked away a free man.

Chapter 23 A Police-Shooting Investigation and the CIA

One evening in a rural part of southern Oregon, four young men drove to a house where an altercation occurred with the local residents. Someone at the residence called 911, and two sheriff's deputies responded to the scene. It was dark, cold and raining, so the deputies decided to set up on the only road leading from the residence back to the highway instead of going to the residence. They would simply wait for the vehicle to arrive and make the stop at that time. When the vehicle arrived, there was a shooting. The driver of the vehicle was injured and taken to the hospital. It was alleged by the officers that the driver attempted to run over one of the deputies with the vehicle as he was leaving the residence. A vehicle was considered a dangerous weapon, and it was appropriate for law enforcement officers to use deadly physical force to protect themselves, if that was the case. The driver was arrested for attempted murder of a police officer.

Through a graphic modification of the actual scene photo (Figure 63), a depiction of the reported event was developed.

Figure 63

The state crime lab had processed the vehicle and filed a report. Judging from the report, the DA believed that the driver intended to run over the officer. As it turned out, based on tire tracks, he never left the roadway. As mentioned before, this was the only way in or out of the residence. Was he trying to hit the officer, or was he just trying to get out of there? Since it was dark and raining, at what point did he actually see the officer in dark clothing? Was his intention to flee or to harm the officer? Exactly where were the two officers?

Gary and I were retained by the defense attorney to look at the facts in the case. Gary was also a retired lab director and a good friend. He had more crime scene experience than I did, but we made a good team.

He had an excellent background in bloodstain pattern analysis and was also an excellent graphic artist. In the mid1980s, he and I used to teach week-long courses on bloodstain-pattern analysis to law enforcement groups. We taught several courses in Oregon and one in Nevada. During the class in Nevada, the group decided they needed a logo, so they had T-shirts made that stated, “My day begins when yours ends.” I suppose it was appropriate for the topic material but crass in the eyes of the general public. Now we were working together on a case, and it was great to have someone to complement your own observations and peer review your report.

We went to the scene and did a search of the vehicle seized after the incident. The scene was in rural southern Oregon on an unimproved road with limited gravel leading to a residence. Reportedly there were no weapons in the possession of the vehicle occupants. We had copies of the medical records, and we observed that the driver was shot in the side of the head, with the bullet coming to rest under the jaw, as seen below of Figure 64.

Bullets had struck the hood four times in front of the driver but had not penetrated the fire wall. There were also two holes in the windshield, one just over the steering wheel. Keep in mind that it was dark, and the vehicle was reportedly moving.

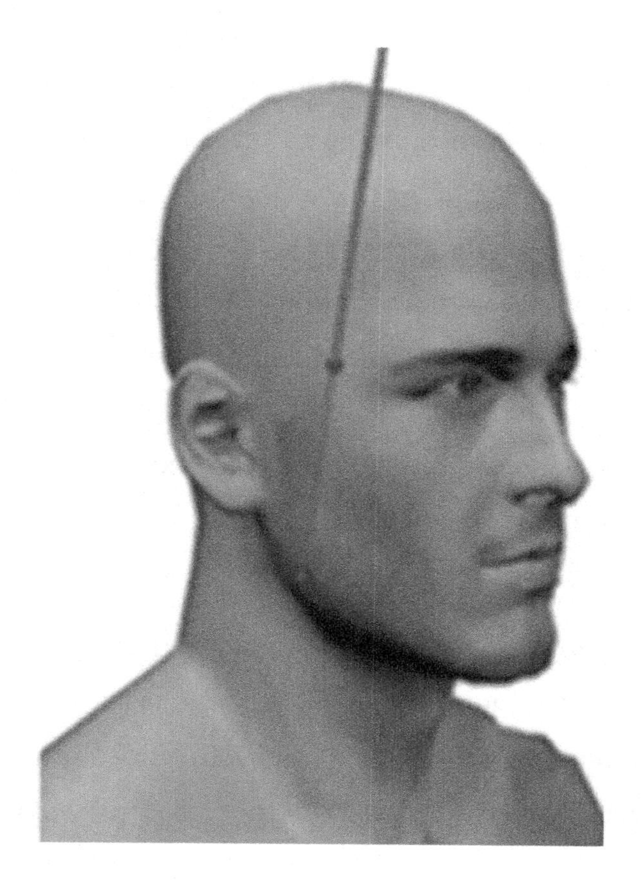

Figure 64

The officers reported that the vehicle accelerated when the occupants determined there were police outside. One officer fired nine times from in front of the vehicle, and another officer to the side fired one round. It was amazing that no other occupants in the vehicle were hit.

Figure 65

In Figure 65 and 66 the rods indicate the bullet entries in a state lab photograph. The following photo shows a diagram on a yellow paper at approximately the height of the driver.

The problem with the photo was that it did not correspond to the injuries of the driver. No matter how we tried to place ourselves behind the wheel, we could not duplicate that bullet trajectory. One could not bend the head forward enough to align with the bullet entries in the windshield.

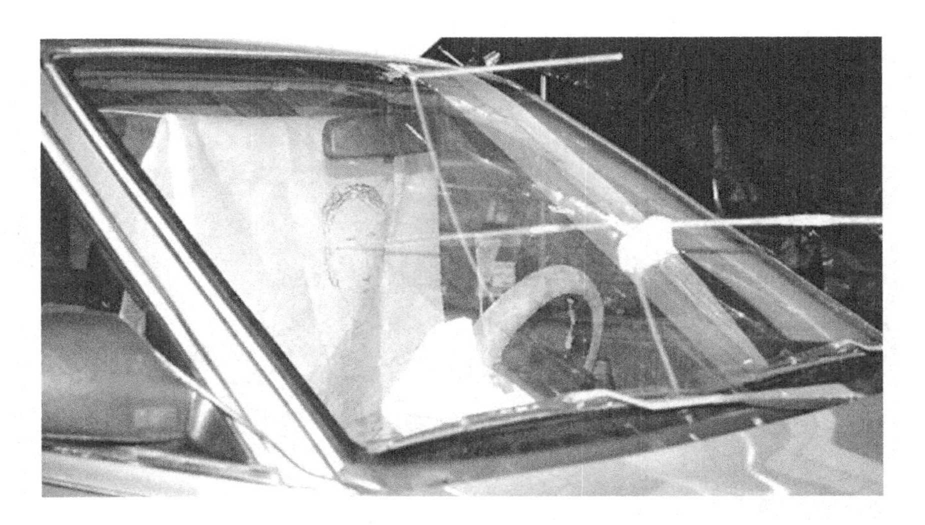

Figure 66

Here the entry was correct, but the path should have continued to the back of the head. The other problem was the use of straight rods, as bullets will often deflect when going through a windshield. There was no mention of this in the reports.

Both officers' weapons were Glocks. Glock firearms utilize a polygonal rifling system instead of the regular lands and grooves found in other weapons. Polygonal rifling is a slightly out-of-round hole in the barrel, and a bullet tends to move from side to side as well as rotate as it goes down the barrel. The result is that often no two bullets from one firearm look the same when you're looking at minute striations under a microscope. The matter gets worse when you have two Glock firearms and you are trying to determine what bullets came from which firearm.

In this case, the bullet that struck the driver was damaged making matters worse. The state's firearms examiner thought that he could tell a difference, in that one of the firearms had been modified (ported) and that left a distinctive mark on the bullet. I had my doubts, so the attorney asked for discovery of his notes. He refused to give them up.

I was not sure if he fully understood the law that defines dual discovery in a criminal case. Was he just being obstinate, or was there something to hide? The one saving grace with Glock firearms was that you could still find individual toolmarks on the cartridge cases that can differentiate the weapons. There were nine casings on the ground in front of the vehicle. From them, we could tell how many shots were fired by each officer.

I set up an appointment, and we went to a local junkyard and asked to shoot a vehicle, we needed to do some of our own testing. I had performed shooting tests there before and they knew that I was careful about how I conducted my testing. The vehicles used were similar to the scene vehicle and held to be crushed for the metal recovery, so shooting the windshields did not matter. We set up a panel behind the steering wheel to determine the amount of deflection created by the bullets. When bullets strike a windshield, the expectation is that it goes straight through; they do not. The bullets will deflect to some degree depending on the bullet type and the caliber of the weapon. Slower-velocity bullets tend to deflect less than faster bullets, such as from a high-powered rifle. This is the opposite of what one would expect. With a Glock model 22c in .40 S&W caliber like what the officers used, I got a deflection angle between two and seven degrees downward. This was still not enough to get a bullet striking the right temple area of the head to lodge below the lower jawbone.

When a cartridge case ejects from a fired weapon, it will cast to the side normally. Test firing a weapon to see where the cases land is called an ejection pattern. One seldom uses ejection patterns for a weapon in order to place a shooter because they are easily moved or bounce when hitting hard objects, thus affecting the reproducibility of the test. But when you have nine in a small group as in this scene, that group can be used to generally place someone based on similar testing. This time it put one officer in the roadway of the oncoming vehicle. This was consistent with the angles of the trajectory rods too.

The cartridge casing from the other officer standing beside the roadway was found inside the vehicle. By the way, the driver's window was down at the time of the incident. The second officer had to be close enough for the casing to end up inside. When a weapon is held horizontally, most semiautomatic weapons eject cartridge cases up and to the right. However, if the weapon is pointed downward, it may eject them forward. That observation seemed plausible for the officer beside the vehicle to have fired the shot that struck the driver as seen in Figure 67.

In our testing, we set up a panel behind the steering wheel and fired a round into the windshield at approximately the same location as the bullet entry in line with the driver of the vehicle.

Figure 67

The bullet penetrated the glass and deflected downward, and the bullet hole was surrounded by small fragments of shattered windshield glass. If a person was close enough to a window when a bullet passed through, the glass fragments would have sufficient energy to penetrate the skin.

Sometimes the glass will create a pattern that looks like an unburned gunpowder pattern, which will lead the examiner to believe it was a near contact shot. There was no glass noted around the victim's entry wound, further support that the officer to the side fired the shot that struck the driver.

This happened in a case in eastern Oregon. The victim was shot through the windshield of his car from a distance of one hundred yards.

The body was shipped to Portland for autopsy, and the medical examiner declared that there was powder around the bullet entry and that that the victim was shot at close range. He did not attempt to identify the penetrating material around the bullet hole, nor did he take a tissue section to identify any burning around the bullet entry. Initially he refused to budge on his conclusion, and it created a problem for trial, as the finding was totally outside the circumstances of the case. Just before trial, the client pled guilty and told all about what had happened. He took the investigators to his location at the time of the shot, and it was as mentioned earlier, it was about one hundred yards out. The medical examiner never did budge on his opinion. Here was one of those circumstances where elimination theory would have been helpful in formulating a conclusion.

Getting back to the testing on the vehicle, our tests showed that the glass dispersion after the bullet was about six inches wide and twenty-five inches long. Enough to strike the driver.

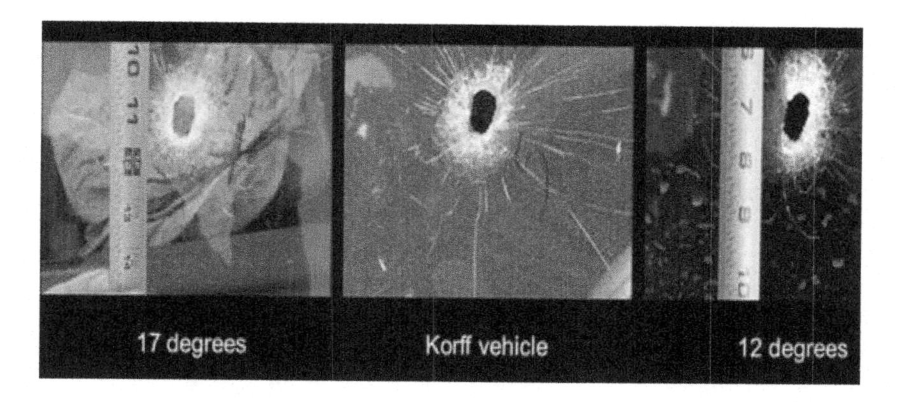

Figure 68

The photograph above in Figure 68 shows the oval entry through the windshield in front of the steering wheel.

At first blush, due to its oblong appearance, one would think the bullet was a ricochet and was tumbling when it struck the windshield, but our tests showed the entry angle to be seventeen degrees. How does this relate to the incident? That was hard to say, as these shootings are seldom static incidents. The vehicle could be moving, or the officer could be moving; or both could be moving at the same time.

The cartridges used by the law enforcement agency had hollow-point bullets, which break up easily when striking a hard object. The driver had only a single bullet entry with no surrounding particulates. The photo in Figure 69 below shows the extent of breakup of the bullet seen in our tests when shooting a single round through the windshield. Each hole in the paper was created by bullet fragments.

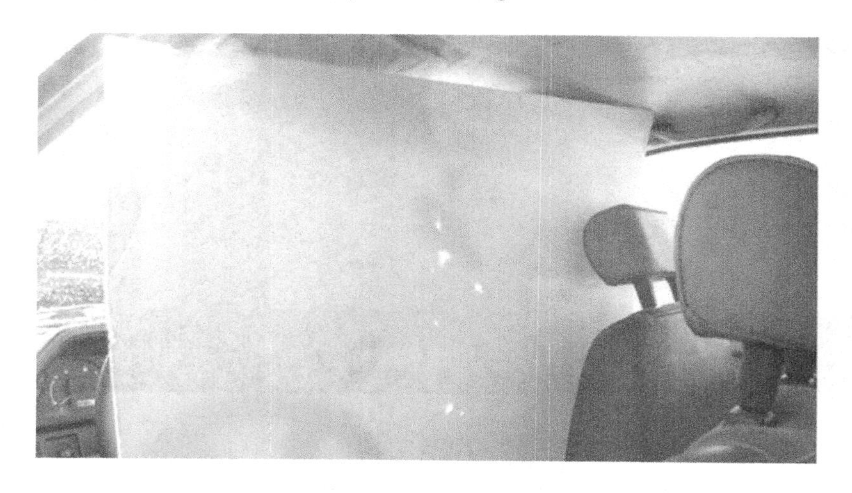

Figure 69

As we reviewed the case and did our testing, we were in contact with the defense attorney and did as much teaching as reporting, but that was not uncommon. To our surprise, the client had a brother who worked in some capacity for the CIA; the brother started contacting the attorney and giving his input on what should be done in the case. He was unusually well versed on our activities, and it was apparent that he had tapped into the e-mails of the attorney and had all the information, including reports on her computer not related to the e-mails. Only recently I discovered I had a software hole on my computer which was a remote access portal. Apparently, the attorney was not the only one hacked. This was a first for me, spying by the CIA in a criminal case.

When Gary and I processed the original vehicle ourselves, we noted bloodstains on the driver's door and driver's seat area. This was a bloodstain pattern that originated between the steering wheel and the door, and it was consistent with back spatter. As mentioned earlier, the window was down at the time of the incident and it was raining. So the blood pattern may be incomplete as noted due to the rain. But despite the potential for incompleteness, the origin could still be determined to be between the steering wheel and the window.

We were concerned that the source of the blood could also be coughed blood from the driver having blood in his mouth after the shooting.

I sent samples of this blood out to a private lab and had it tested for saliva and nonnucleated epithelial cells. These cells are common to the inside of the mouth and are shed in large quantities. The tests came back negative for both. This does not mean that it was not possible, only that the tests did not find any indication of saliva.

Figure 70 represents the bullet removed from the victim. On microscopic examination, no glass was embedded in the bullet. All of the test fired bullets in our study that passed through a windshield has minute glass fragments embedded in the lead.

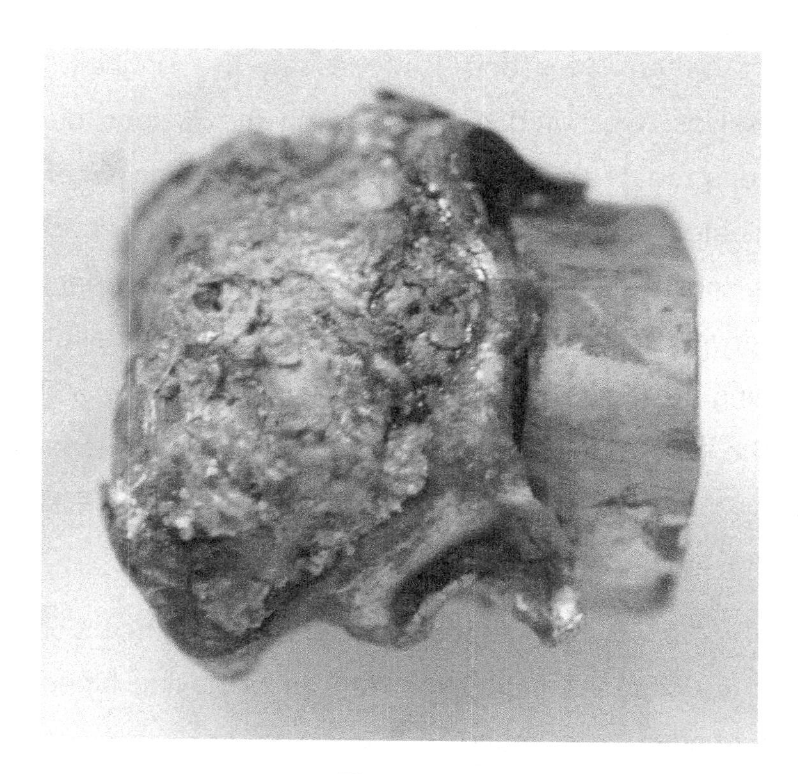

Figure 70

When I went back to the reports and checked, the bloodstains on the door were photographed by the state lab, but there was no mention of them in their report. Did they forget?

As a crime-scene expert, one had to keep in mind that the photograph is a compilation of all activities that took place between the incident and the time the photo was taken. Beware of assumptions that the photo depicts only immediate post incident activity. After the driver was shot, it was unknown what bloodletting took place before he was removed from the vehicle. As I recall, he was conscious, and the officers removed everyone from the vehicle for safety reasons, as was common police protocol.

Focusing now on the second officer beside the vehicle, our testing demonstrated that if the pistol was held at a forty-five-degree angle, the expended cartridge case would travel forward at sufficient height to enter the driver's window if the window was close enough. Examination of this casing inside the vehicle proved the source as this officer's weapon. The photograph below in Figure 71 from the scene demonstrates how the vehicle was tilted, which could have facilitated the casing from the second officer entering the driver's window.

In an additional step, I requested the bullet recovered from the driver and specifically looked for glass fragments with a microscope. None was visible.

In the calculations associated with the holes through the windshield and the hood, it was apparent that the officer was not in front of the vehicle at the time he was shooting but off to the left (passenger) side. It was also apparent, through the process of elimination, that the second officer beside the vehicle had fired the shot that struck the driver while he had his head down between the steering wheel and the window.

Figure 71

If the driver had his head down to avoid shots through the windshield and the vehicle was still moving, the movement of the vehicle may have been erratic.

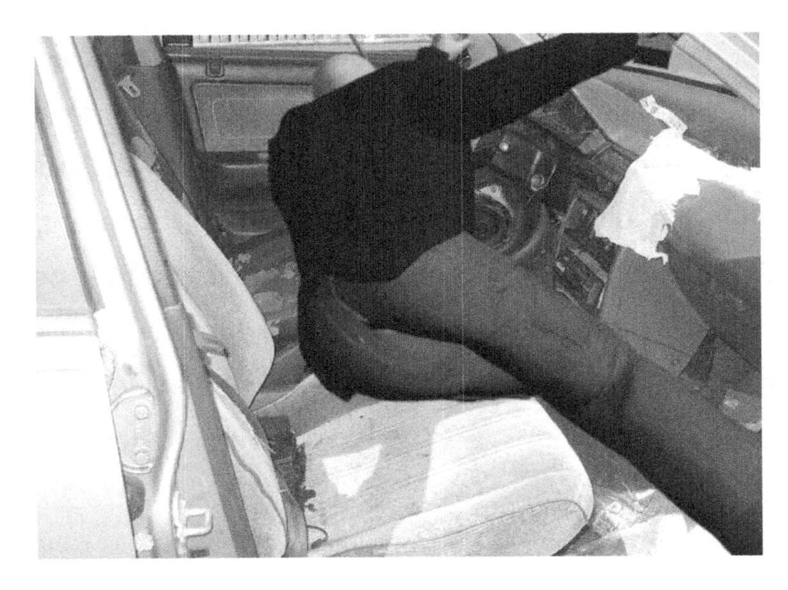

Figure 72

The graphic above in Figure 72 depicts a possible position of the driver at the time he was shot. This was based on the bullet trajectory within his head and the shot originating from the open driver's window.

At trial, I gave a PowerPoint presentation of our findings and indicated that both the officers were not in front of the vehicle as described and the source of the charges against our client. The state put on the crime-lab firearms examiner, and the defense attorney asked for his notes. He said, "No, I will not give them up."

Gary and I looked at each other like, "Are you kidding me? Doesn't he understand the discovery laws?"

The attorney appealed to the judge, and the judge swiftly instructed him to turn them over. The attorney asked to make a duplicate copy of the notes, and the examiner said that he did not want the notes out of his sight. Again we looked at each other. What did he expect we were going to do with them?

It was getting close to lunch, so the judge allowed the examiner to follow us over to the defense attorney's office so that he could watch his notes be copied. Gary and I found all of this to be amusing, as the notes really did not have anything new to offer. When the examiner got back on the stand, he took a few shots at our work, but no harm was done. He had been to the same firearms-shooting class as I had, so there was nothing out of the ordinary.

The crime scene itself was processed by two other crime lab people who had reviewed our discovery and reportedly one had agreed with our conclusions. When we learned this, we encouraged the attorney to have him subpoenaed for our case. Our attorney said no; she would address this issue when he was on the stand. As it turned out, in a stellar move, the prosecution did not call him. Our attorney called the lab to get him to testify, but wouldn't you know, he was on vacation out of the country. A lesson learned.

He made the statement that the deputy was in front of the vehicle in a kneeling position perhaps, or did he shoot from the hip? We did not dispute the officer's location was unknown prior to the shooting.

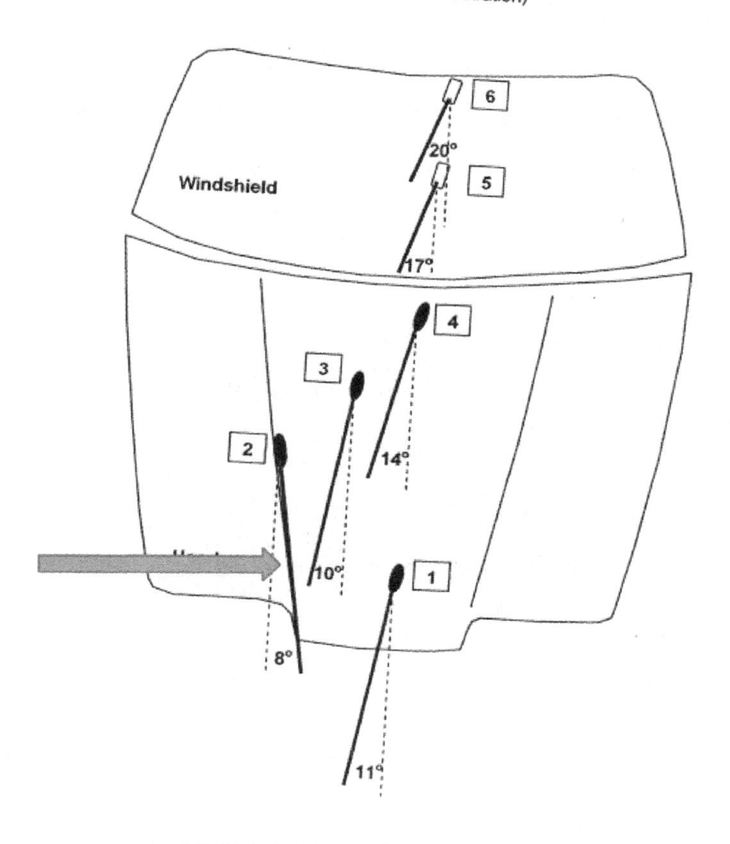

Figure 73

But he did manage to get to the side, and the trajectories shown in the diagram above reflect that. We also demonstrated evidence that the driver was shot by the deputy beside the vehicle, not in front.

The text from the crime lab report above was a concern, in that the movement of the deputy singularly could have caused the spread in the bullet entries in the vehicle. It did not take into account any sudden movement of the vehicle not seen in the available tire tracks. Ever ride in a vehicle down a bumpy road going too fast? The tire tracks may not reflect what the rest of the vehicle is doing. If you looked at photos of the scene, there were areas of water puddles were no tracks were visible.

Last, I don't know a civilian, let alone a police officer, who would stand in front of an oncoming vehicle full of known combatants who had recently committed a crime on a dark and raining night out in a rural setting. That was a concern.

It is speculating at this point, but the scenario could also have gone like the following. The officers had prepared for the oncoming vehicle and had set up on opposite sides of the roadway. The problem was cross fire; if the shooting started, they would be shooting at each other. It could have been that one officer decided to change sides of the roadway just before the oncoming vehicle got there and attempted to run across the road in the headlights of the oncoming vehicle.

Problem could also be that he slipped and fell. Now you have a crisis. This situation was dangerous, and the officer started shooting as he got up and got out of the way. The pattern of shots from right to left seen on the hood and windshield would also fit this scenario as the officer was running to the left.

How about this? The officer was not in front of the vehicle but to the left of the oncoming vehicle and stood in one spot. As the vehicle approached and he started shooting, the angle between the long axis of the vehicle and the bullet source increased. The second shot mentioned above may be from a greater distance as the vehicle attempted to maneuver down the road. These are scenarios that cannot be eliminated on scientific information only.

If the officer was capable of getting out of the way, where was the necessity to neutralize a perceived deadly threat? Perhaps it was just a split-second judgment call that cannot ever be dissected to the satisfaction of all parties. I have investigated numerous police-officer-involved shootings in my career and demonstrated alternative choices, as in this case.. A lengthy investigation can often sort out facts unknown to the scientist. Sometimes the attorney on either side does not want to give you all the information for fear you may change your opinion. In this case, you can see the complexity of the problems in seeking the truth. Very limited facts were generated prior to the initiation of shots.

In spite of alternative scenarios, in the number of police involved shootings I have either worked directly or reviewed for the defense, my money is on the police 99 percent of the time.

That 1 percent is what kept me vigilant.

During the trial, I was able to hear the testimony and I kept waiting for the officers to state at which point the vehicle turned out of the roadway toward them. I never heard that statement, but perhaps I missed it.

Our client was not a pillar of the community, and although we were able to demonstrate that the officers may have moved to the side of the vehicle, he was still convicted of the crime.

Not long after the trial ended, I got a call from the brother thanking me for all the work I had done. He had been able to keep up with all the developments from Washington, DC, through reading the reports, e-mails, and other documents. Who knows who else he hacked. I thanked him and did not inquire about how he knew so much about the case. I just moved on. My one and only encounter with the CIA. Or maybe not. Who can say for sure?

Chapter 24 The Multiple-Shot Self-Defense Case

In April 2006, a young man named Matthew was rooming with another fellow in an apartment in east Portland. He was a college graduate in music but was five inches shorter than his roommate and forty to fifty pounds lighter. Matthew had taken an NRA firearms class and liked to shoot. But most of all, he liked rap music and just liked to hang out with friends. The roommate had a felony conviction, so he had Matthew carry his gun for him.

In this dynamic, the roommate was the dominate male. He liked to drink, got abusive when he drank, and liked to be called Kadaffi. This subservient relationship was interesting in that it was abusive but reportedly not homosexual. One witness described the relationship in a condescending manner as "like the way you would treat a girlfriend that you did not care much about." Kadaffi did not pay any rent or buy any of the food. When he sobered up, he would apologize, and things would be back to normal. A common theme in abusive domestic relationships.

One afternoon when Kadaffi was drunk and in a rage, he began beating up Matthew and stated that he was going to kill him. The assault became more violent until Matthew finally feared for his life and pulled out the pistol and shot Kadaffi five times. Matthew called 911, and the police arrived to do an investigation. When interviewed, Matthew stated that Kadaffi was in a rage, was on top of him, and wasn't making sense of anything. He felt threatened and did what he did while fearing for his life. The interviewer did not like the multiple-shots idea, and in time, Matthew was arrested for the murder of Kadaffi. This was when I got the call from Steve Lindsey, his attorney.

Steve and I went to the apartment and did a search long after the police had left. The apartment was nicely furnished with two televisions, a couple video-game machines, and decent furniture, and it was clean. All we found that was out of place was a bullet hole in the wall.

It was five foot ten inches above the floor and had an acute angle toward the floor. Defense attorneys can be useful. In the photo of Figure 74 above, I put him on the floor to show the bullet trajectory. At this point we were not sure how this fit in the incident. We knew the clothing on the deceased was baggy, in tune with the clothing fad at that time. There were seven shots fired, and he had been hit five times. All of the shots to the body demonstrated stippling from gunpowder. The knuckles of the deceased were bruised.

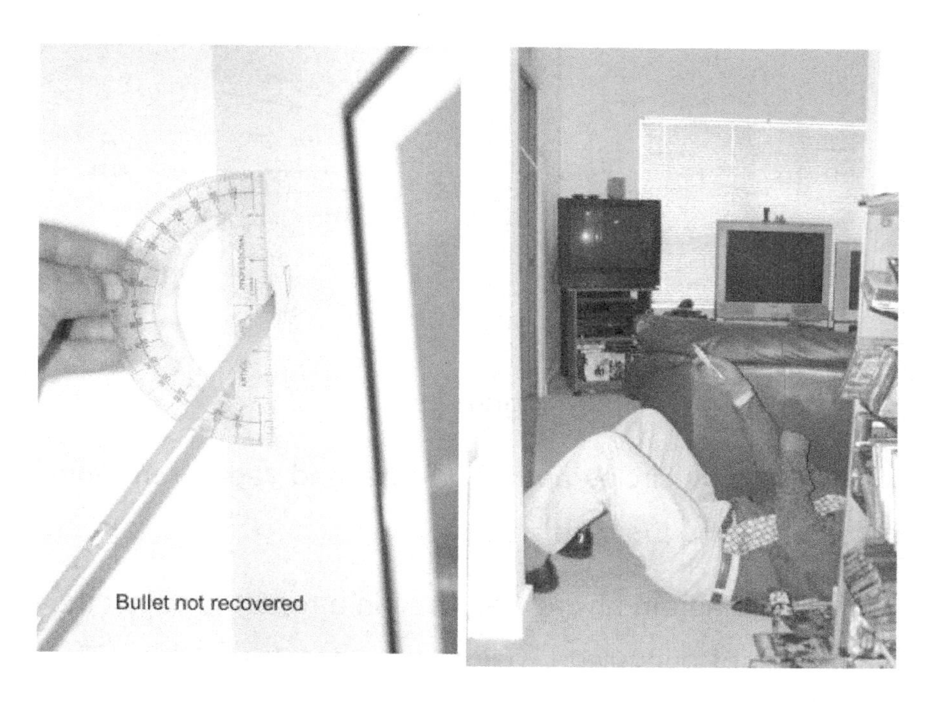

Figure 74

As a side note, after the bullet exits the barrel, often there is unburned or partially burned gunpowder that follows. The shorter the barrel, the more particulates are produced.

If the target is close enough, these particulates have enough energy to penetrate the skin. *Stippling*, as it is called, can be seen as the red dots on the skin around the bullet hole as seen below. These are caused by the gunpowder.

The larger the pattern, the greater the distance. If you have the original firearm and ammunition, the diameter of the stippling can provide a specific distance from the muzzle to the target represented in Figure 75.

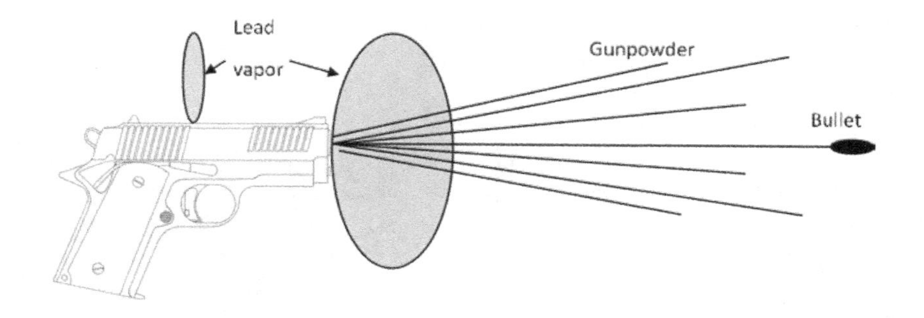

Figure 75

The gray circles above indicate lead vapor from both the receiver and the bore.

The diagram above was used in a report to depict the distribution of gunpowder particulates and lead vapor. The result of the particulate distribution can be seen in the photo below.

From the entry wound, shown in Figure 76, the path of the bullet becomes important and was determined at the autopsy in order to reconstruct the incident. The bullet from the entry shown above was recovered from the upper arm. At this point I have done nothing to confirm or deny the statements of the accused.

The deceased had five bullet holes in him, and all of them would be important to determining why so many shots were fired and how could this be self-defense.

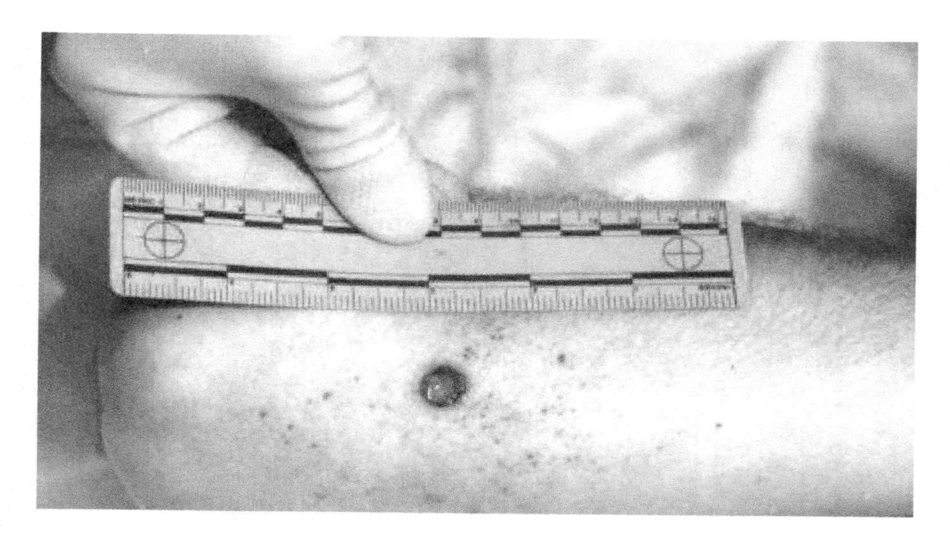

Figure 76

It is important to note that only shots to the head or spine will stop action immediately; some continued action is possible with shots to other locations.

In the diagrams below, numbers reflect the bullet-entry numbers in the autopsy report, not necessarily the sequence of the shots, and there may be alternative sequences. The graphic below would depict a compatible shot to the right arm or shot number one in the diagram.

In our graphics, such as figure 77, the apparel is similar to what was worn by the combatants at the time, and the posters on the wall were there at the scene. The size of the two individuals is set in the software in order to be reflective of the individuals.

Figure 77

When graphics are used in a report or trial, I always caution the reader or listener that the graphics are approximations designed to help the reviewer understand my opinion. They are never to be viewed as an absolute fact.

In the graphic above, the distance from the weapon muzzle to the arm and the path of the bullet was accurate. The rest was subject to change, based on statements of the witnesses or client.

It was not uncommon to have the graphic artist in the courtroom at the time of testimony. In the event objections were voiced, changes could be made prior to the jury seeing the graphic.

Figure 78

This photo of Figure 78, Kadaffi's shirt, showed the bullet hole in the upper left torso, which was also indicative of a close-range shot. The graphic depicts the potential position of the combatants when the shot number two was fired. Figure 79 might represent a shooting position for the combatants.

There are chemical tests used to determine the presence of lead and present a purple color. Those tests were performed on Kadaffi's shirt by the OSP Crime Lab. One area near the right hip area, seen in Figure 80 of the shirt was a problem, in that there was no bullet hole in the shirt to accompany the lead-vapor deposition.

Figure 79

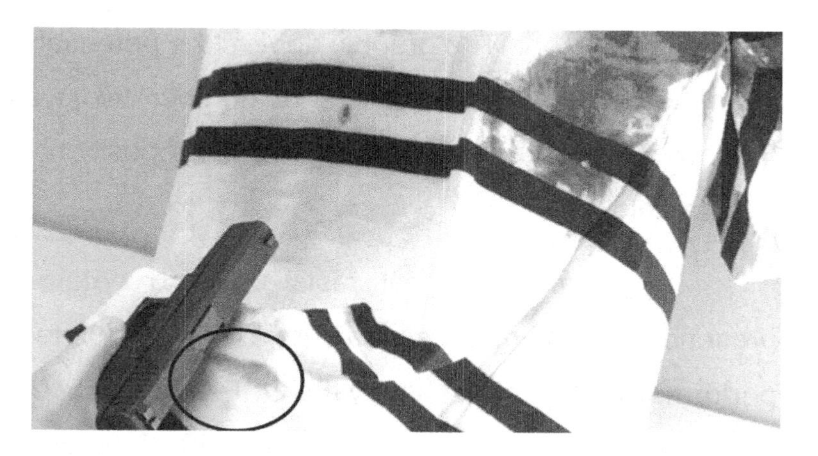

Figure 80

When we placed the weapon against the shirt while on a mannequin, the relationship was quite clear. The lead vapor had escaped from the chamber. The shot might be depicted in the graphic 81 below. Was that the position of the right arm?

Maybe.

Figure 81

Getting back to the crime scene, remember there was a hole in the wall with an acute downward angle. How did that relate to the altercation? On the shirt of the deceased, there was an area on the upper left sleeve that was positive for vaporous lead, but again there was no bullet hole. This may have been a miss. The photo below in Figure 82 of the shirt shows the purple area on the sleeve from the chemical testing.

Figure 82

If you use the client's statements and orient the combatants appropriately, the graphic below in Figure 83 showed how the hole in the wall would demonstrate the miss. Could the sleeve be in another position? Maybe. Before the graphics were generated, I sat down and talked with Matthew about what he could recall. The graphics were generated and later shown to him to determine if they reflected the incident. He said they did.

Figure 83

In a review of a case like this, the first problem to overcome was that the photos of the scene were static. An altercation is fluid and usually is over within seconds or minutes. In addition, the memory of the witnesses or the client can be clouded with the emotions during the event, moments of blackout, or constant reevaluation of what happened later. These issues could change the picture of what really did happen. In my experience, they are not lying; it is just that the circumstances can mess with the mind. The physical evidence becomes very important in order to back up memories or point out inconsistencies.

Once the decision to shoot is made, all shots are usually in rapid succession. The struggle may continue until the blood stops flowing to the brain and the person blacks out before finally passing on. Sometimes this could take minutes. Was the shooting justified? It appeared that Matthew was on the bottom and feared for his life. That meant that he could use equal or greater force against his attacker to save his life, according to the law.

The trial was set before a judge and no jury. The state began with opening arguments and questioning the police and other witnesses favorable to their case. Last were the state crime-lab people, who made no attempt to reconstruct the events in this case. They only mentioned the examination of the weapon and the clothing. They did the proximity testing on Kadaffi's shirt, so we had no complaints about that. When the state rested, the defense put on a pathologist, a firearms examiner, the graphic artist, and I was last.

The case took all day with several recesses between witnesses. During a recess, we overheard the prosecutor say that she was prepared to cross-examine me and expected to make the most of it. My attitude was, bring it on. But the defense attorney had a new idea. All the information about the deceased, especially the bullet trajectories, came in through the pathologist. He was also asked if he had seen my report and the graphics. He stated that they were accurate from an anatomical viewpoint.

The firearms examiner was handed my report on the stand and asked if the proximity testing and the distance of the weapons in the graphics was reasonable and accurate. He said yes. The graphic artist was next.

Unlike most graphic artists, Gary was a forensic scientist and the primary partner in my business. When he starts knocking out renderings, you can bet they are the best available. More reconstruction information came in through him. It was clear to everyone that the prosecutor was just letting everyone through while waiting to take a shot at me. At times she did not cross examine on some of the most basic issues. In fact, all of the information in my report was now into evidence through the other experts. When it was my turn, I could see the prosecutor gearing up with some excitement in her eye. Then it happened…The defense rested! They got all they needed into the record; I did not need to testify.

The prosecutor had a whole host of questions to ask and was saving them for me. She never got to get them into the record. And since she knew I would be in the courtroom, she did not subpoena me, either. Wouldn't you know it. I left the courtroom just minutes before I was scheduled to testify and was unavailable.

The judge ruled in favor of the defense, and Matthew was acquitted of all the counts against him. In the last sixteen years of my career, I worked primarily for the defense.

There were only a handful of cases where a person accused of murder was acquitted based on my scientific examinations. This was one of them. Matthew walked away a free man.

Chapter 25 Homicide or Suicide

In the fall of 2009, I received a call to examine a scene, clothing, and a firearm in a shooting in California. The scene was a clean one-bedroom apartment with the usual decor. The bedroom had a queen-size bed in the middle of the room. This made the distance between the entry door and the end of bed just enough space to get the door open.

I was told the victim's husband was home at the time and heard a shot. He reportedly rushed upstairs to the bedroom and found her on the floor partially blocking the door. To gain entry, he had to move her. He stated that when in the room, he picked up the .22-caliber semiautomatic rifle off the floor and placed it on the bed. The weapon was a nylon stocked Remington model 66. The overall length of the weapon was about thirty-eight inches, an important fact for reconstruction. The husband was also known to have alcohol issues, but I was unable to locate a specific blood alcohol value for him. He called 911 and reported his wife had been shot, and it was during this conversation with dispatch that he picked up the rifle off the floor and from under his wife and placed it on the bed.

In the many years that I taught crime scene investigation, I often made reference to the Pex rule:

If it is a witnessed death, the case is treated as a homicide until you can prove it is not.

His wife was on the floor in their bedroom with a bullet entry in her forehead. Despite the shot, she was still alive when medical personnel arrived and transported her to the hospital. She died the next day. Her blood-alcohol level was about four times the legal limit; plus, other prescription substances were in her system. Uniformed officers were first on scene, followed by EMTs, followed by detectives, who would process the scene. The detectives decided there was far too much bloodspatter for a suicide attempt and arrested the husband for murder.

When interviewed, the husband was still under the influence of alcohol, and when asked if he had shot his wife, he said that he simply had no recollection of doing that but perhaps he had. Although I was not provided with toxicology data from blood and urine testing, there could be a reason that these tests were not done on the husband. I have witnessed circumstances where a suspect in a crime was not tested for drugs or alcohol because the prosecution did not want to give the defense an easy out at trial. It was my observation that the old defense of "*my client was so out of his mind that he had no idea what he was doing,*" usually will not get someone off, but it could get the charges reduced from murder to manslaughter or a reduction in incarceration because of the mitigating circumstances. There is normally no legal obligation for the prosecution to test the accused for drugs or alcohol.

The prosecution employed a bloodstain pattern expert to review the evidence. He noted the bloodspatter at the scene, on the firearm and on a sweat shirt reportedly worn by the husband. On examination of the sweat shirt, he concluded there was impact bloodspatter consistent with originating from a gunshot. The old term for this was *high velocity impact spatter*. The conclusion bolstered the state's case and the husband remained in jail. The expert also supported the fact that there was too much bloodspatter on and around the bed for her shooting herself.

This information must have been passed on to the medical examiner who did the autopsy, because the cause of death listed in his report was homicide. The ME described the bullet entry as a contact shot with the muzzle of the rifle against the forehead. In looking at the photos from the autopsy, one could see the red rim of abrasion created by contact with the muzzle of the firearm as shown in Figure 84. This rim was about one-half inch in diameter. The diameter at the muzzle of the rifle was nine-sixteenths of an inch, which was close enough, as the muzzle is rounded; the outside edge is not the leading edge. The center of the wound was black from the burning gunpowder, what we call *sooting*.

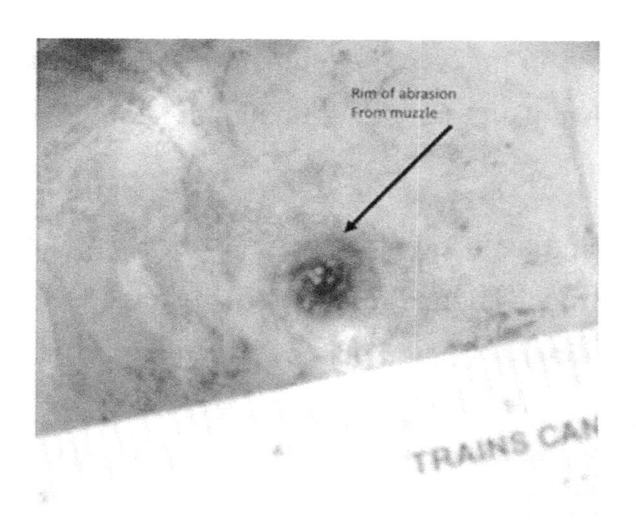

Figure 84

On the firearm side, bloodspatter leaving the wound will land on all adjacent objects that are in a direct line. This includes inside the bore of the weapon and on the hands and clothing of the shooter as well. The first expert thought that this was the source of the blood on the sweat shirt. Since this spatter, known as *back spatter*, was very small and only travels a few feet, this would place the husband wearing the sweat shirt within two feet of the wife at the time of the weapon discharge. The implication was that he was the shooter.

An attorney was hired to represent the defendant, who also hired an expert to review the evidence. This expert reached an opposite conclusion regarding the blood on the sweat shirt. He concluded that the blood was a transfer pattern, not the result of gunshot. In all the years I have been in this business, the usual remedy is to bring in the experts at trial and let the jury sort it out. It is a dangerous approach that may rely more on the ability of the experts or the attorneys to express themselves rather than the facts regarding the evidence. This time the prosecutor and defense attorney discussed the findings and decided that there needed to be a third independent review. It had to be someone whom both experts believed to be reliable, or it was a no-go. I don't know how the discussion went or how my name came up, but I got the call. Apparently, the understanding was that the third opinion would be the deciding and final opinion. Wow!

When I got the call from the prosecutor and he explained the situation, I could not believe my ears. From a science standpoint, this was commonplace, but from a legal perspective, this was unprecedented. I asked who was going to pay, and I was told that the bill would be split between the prosecution and the defense. The accused was charged with murder, was still in jail, and had been for over a year. I was sent the scene reports and photographs, the autopsy photos, report, and some of the interview statements of the accused. The agency refused to send me the evidence but would allow me to see it at their facility.

When a rifle was used in a suicide, the first question was, could the person reach the trigger and get the bullet trajectory described at autopsy? If it was not possible, there was no sense looking at the rest of the case. The only way to determine this is to find someone of similar size, put the same rifle in that person's hands, and see if he or she can reach the trigger. My wife was that person, and yes, it was possible. I kept telling her that the gun was unloaded but had to show her anyway. I can see her point. Was it the chance for the perfect crime? I would only have to say, "Honey, I did not realize it was loaded." I think I needed to work on that marital-trust thing some more before we looked at another case. It was also reminiscent of my old lab days with OSP, seeking volunteers for my experiments when people would ask, "You want to what?"

In preparation for travel, I asked if I could interview the first officers on scene before I looked at the evidence. None of the reports addressed their observations at the time they entered the scene.

Days later I was in southern California at the agency. The people who were there to greet me were unsure of what to expect, and the gathering was quite formal. After all, who was I, and what crime-scene expert did interviews? Cameras were set up. The prosecutor and his staff were present in the room when the first officer on scene came in for us to talk.

Talk about awkward. I had the feeling that I was under the microscope and we were not getting off on the right foot. I kept telling myself that I was not the enemy. I introduced myself and asked the officer to go over his actions and observations when he entered the apartment.

His statements were in line with all the statements in the other reports until I asked, "What was the husband wearing when you entered the apartment and encountered him?"

> The response was, "T-shirt and jeans." The hushed tones in the room screamed, "Wrong answer!"

The officer also caught the change in tone and was looking nervous and glancing around the room. But that was what he remembered. Keep in mind, the sweat shirt in question in this case had to be on the husband as proof that he had shot her. It was removed from him when he was in custody with no mention that he had put it back on after the incident. The interview quickly concluded, and the officer was escorted out of the room by other officers who were wearing frowns on their faces. Here is another example of eye-witness observations; it is hard to get every detail right if you do not write the details down at the time. Or was he right?

Why was bloodspatter so important, and what were the possibilities for its origin?

In this case, we were considering transfer stains where blood was transferred from a bloody object to a non-bloodstained surface. Or it could be impact spatter, where energy was applied to a source and blood was forced into flight to a final destination. A gunshot has the energy to create impact spatter under certain conditions.

When a bullet leaves the barrel of a firearm and enters the body, there are two responses. First, the bullet pushes against and penetrates elastic tissue and creates a wound channel that is larger than the diameter of the bullet. After the initial compression, the tissue will rebound back into place to its original position. Some blood in the bullet track is pushed back toward the entry hole and out into the environment. The second factor is the action created by the expanding gases exiting the barrel behind the bullet. In the discharge of a weapon, there was gunpowder present inside a cartridge. Upon ignition of the powder, the solid powder instantly turned into a hot gas, which expanded, forcing the bullet down the barrel. If the muzzle was against the body, the expanding gases exited the barrel, entered the body, and continued to expand the wound channel. The gas cooled rapidly. Then it collapsed in volume inside the body, and the tissues rebounded to or near their original position. Any blood that got into the wound channel was expelled rapidly back toward the weapon and the shooter.

This was the theory about how the blood got on the sweat shirt. In the photo below, the spatter in question can be seen on the back side of the left sleeve. Bloodspatter does not travel around corners. Take the time, and act like you are holding a rifle in any shooting position you can imagine. Each time look at this area on your left arm to determine if it is a straight line from the end of the barrel. The blood is between the wrist and the elbow on the underside. Keep in mind that cloth will move and stretch slightly when worn.

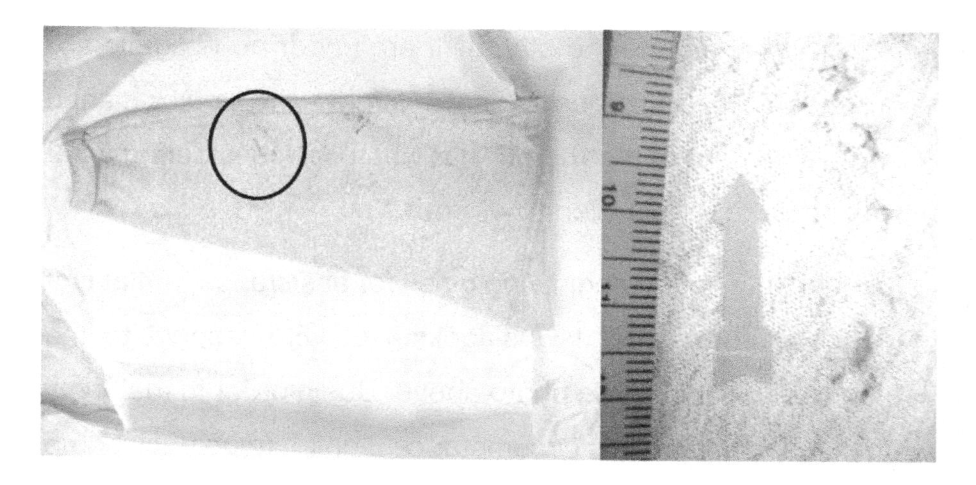

Figure 85

The arrow in the previous photo, Figure 85, was near the bloodstains in dispute. One expert thought they were impact spatter; the other thought they were transfer. With transferred bloodstains, the cloth has to lightly touch a liquid bloodstain to transfer some of the blood.

The bloodstains seen in the scene photographs were larger than what was on the sweat shirt, so it was probable that only a small portion of the original stain was transferred. How was that possible?

I finished up the interview and moved to the small laboratory inside the police station. The small lab was full of people from both sides of the case and a property person, who would open the selected evidence for my viewing. All seemed cordial and interested, except the defense attorney. In front of everyone, he stated, "I am under no obligation to accept Mr. Pex's findings as final." He gave me a quick glance and walked out of the room. That was odd, considering what was at stake for his client.

I only had a magnifying glass for assistance, but at first glance, I knew what I was looking at with respect to the bloodstains seen in the photo above. But in order to make my point, especially in this circumstance, I opted to go back home and do some experimentation before offering an opinion. Being able to reproduce a pattern by a specific action was the best evidence that the truth had been confirmed.

One item of evidence that was not mentioned in verbal discussions was the T-shirt worn by the victim. In the photo below, I could see impact spatter on the top of the left sleeve. In a photograph, it looked like the spatter was on the back, but when placed on a mannequin, the spatter was actually on top of the arm. This would be important later in the conclusions and graphics attempting to develop the position of the victim at the time of the shooting.

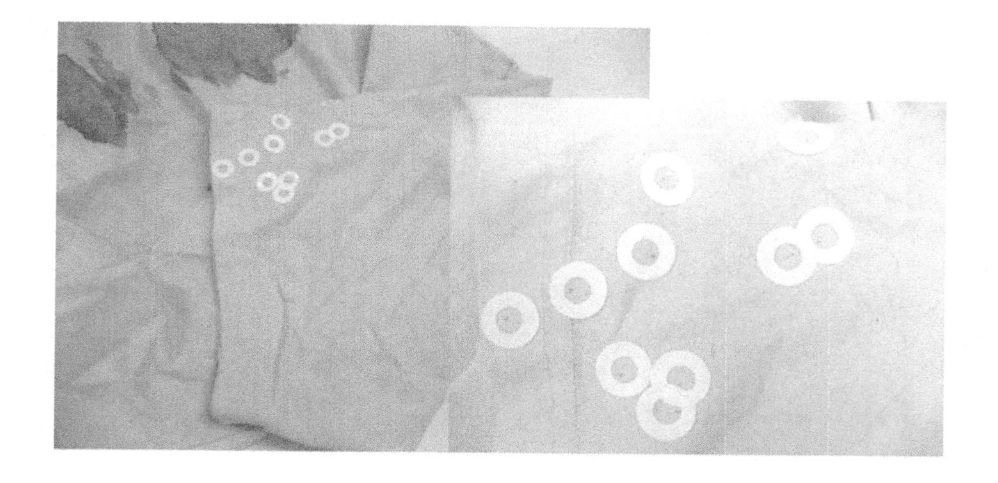

Figure 86

In addition to the impact spatter on the sleeve, Figure 86, there was a transfer pattern on the front of the shirt. It was circular and very close to the diameter of the end of the muzzle of the .22 rifle. The circular transfer pattern was shown in

Figure 87 on the victim's shirt. I attempted to duplicate the mark with another rifle, and the result was demonstrated below in Figure 88. The transfer pattern on the victim's shirt was consistent with a muzzle imprint. Now consider the implications, either the husband poked her with the end of the muzzle or the muzzle touched her here as she fell to the floor.

Both scenarios were possible.

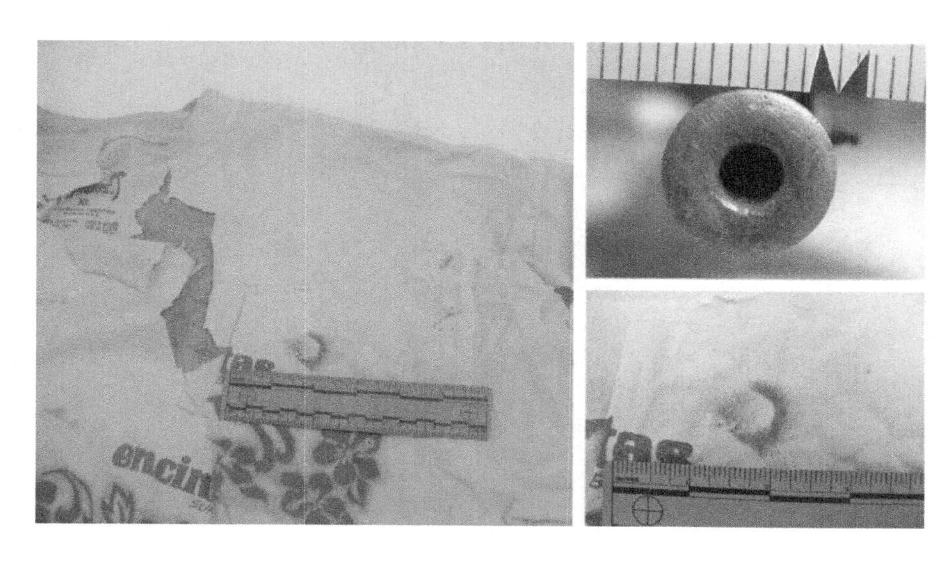

Figure 87

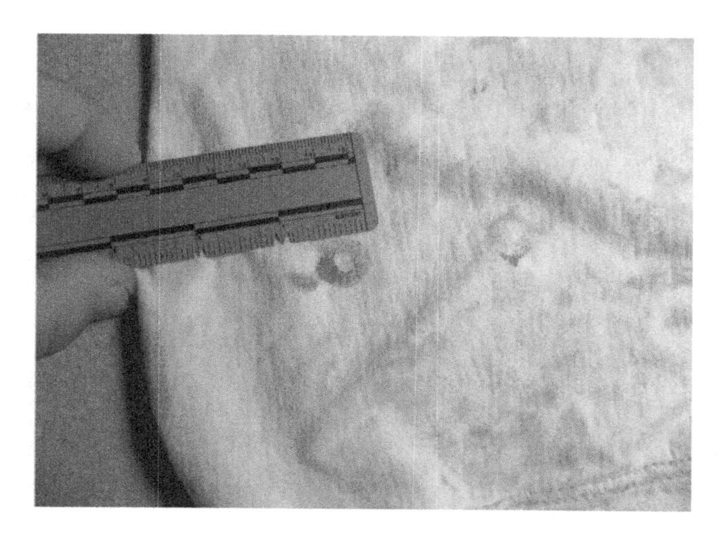

Figure 88

Back to the sweat shirt worn by the husband. In order to make my hunch work, I needed an old sweat shirt, and there was no better place to find one than the local Goodwill store.

There I found several, but I was looking for the same material in a light color to improve the subsequent photography. The test was simple; place blood on a nonabsorbent surface, and lightly touch the sleeve of the old sweat shirt. Note that the blood only touches the fiber tops, which is typical of a transfer pattern. Initially I was not getting the pattern to duplicate the sweater from the husband. The next day I rubbed the bloodstained area together and produced the appropriate pattern on the right. Keep in mind that the sweat shirt had been handled by several people and stored in a bundled-up manner for a year. The final pattern observed on the husband's sweat shirt was result of transfer combined with additional movement or handling after the blood was dry. The test results are shown below in figure 89 and compared to the evidence sweatshirt.

For the final explanation, I needed to demonstrate an impact pattern in order to show the difference to all parties concerned. To produce an impact-spatter pattern, I put on the sweat shirt, placed a blood-filled sponge in front of a bullet trap, and shot it with a .22 rifle. The pattern resulted in small blood droplets striking the cloth random to the weave (Figure 90). Some spatter struck the top of the fibers; others were deep in the weave.

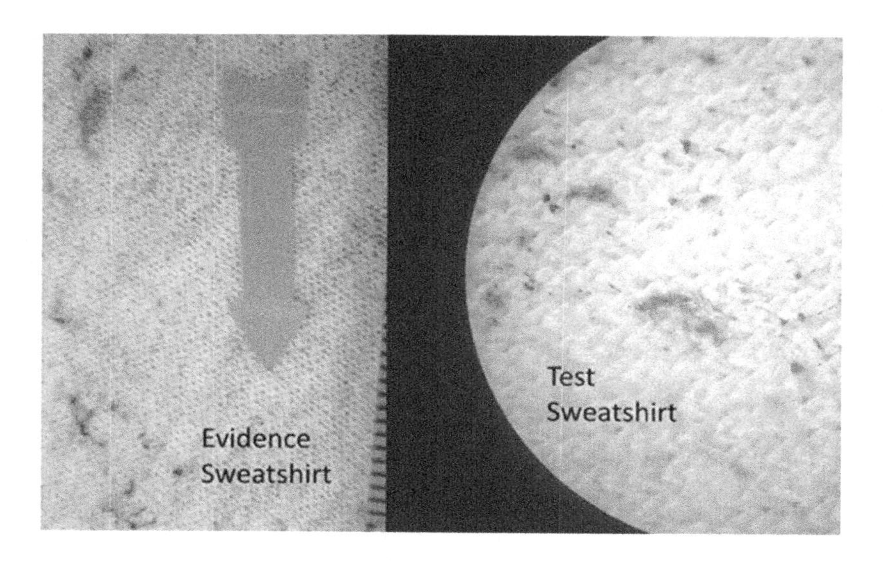

Figure 89

Figure 90

The pattern seen on the husband's sweat shirt did not match an impact-spatter pattern. Unfortunately, having to show the photos in black and white makes it difficult to show the bloodstains.

The rifle also offered good information on what transpired. Recall that the husband reportedly picked the rifle up off the floor from under his wife and placed it on the bed shortly after he heard the shot. In some areas, the blood on the weapon might still have been wet and would smear or transfer. If his version of the events was true, there should be spatter from the original shot on the weapon or from continued loss of blood from the wound.

What I needed to find was original spatter undisturbed by the handling. If the police were correct and he shot his wife and then placed the weapon on the bed, the physical dynamics of the bloodspatter should be different from what I was seeing. The shaded areas in Figure 91 are where impact spatter might be seen depending on if she shot herself or was shot by her husband. Keep in mind that she was wearing a short-sleeved shirt despite the graphic shows a long-sleeved shirt. This means that these areas are physically available to receive spatter in the suggested scenarios.

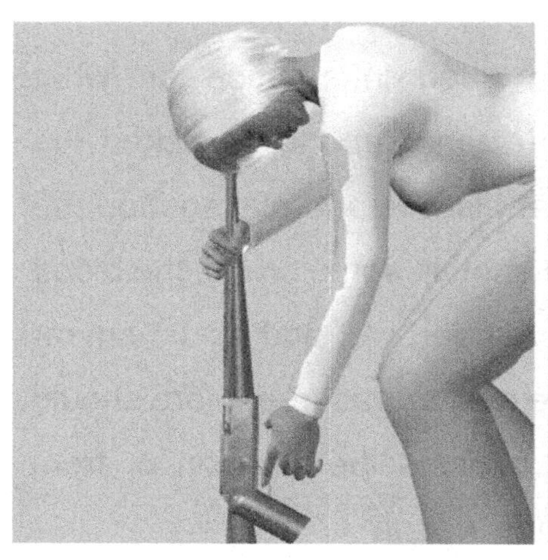 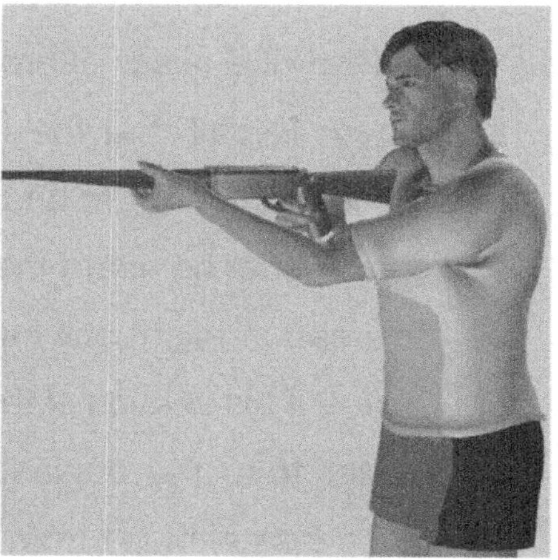

Figure 91

In the photograph below, the first obvious bloodstain was seen on the stock behind where the hand would be if the husband shot the rifle. The blood traveled in the direction indicated by the arrow, so the origin was toward the muzzle. The hand could not be present.

Gravity is the most important factor here and the long bloodstain indicated a low angle to the weapon. The weapon had to be vertical at the time.

The second bloodspatter information to add relevance to this case was bloodspatter on the checkering of the handle, Figure 93. This was where someone would normally hold a weapon when shooting, as shown in the male graphic above.

Figure 92

If the area was covered by the hand, bloodspatter could not have got there. Again, in conversion from color to B&W for this book, it was difficult to demonstrate some of the spatter that I saw on the rifle. Red on brown does not offer much contrast in a B&W photo.

In cases like this one, the opinion of what happened is based on the combination of facts, not just one piece of evidence. There may be other possibilities, should new information become available.

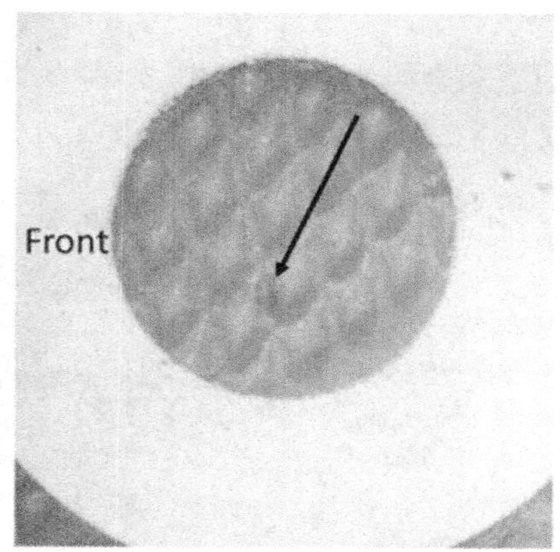

Figure 93

The following was the conclusion I wrote in my report for this case:

<u>*Conclusion*</u>

The blood patterns observed on the rifle stock of the Nylon 66 originated from two directions. The pattern on the buttstock reflects a drip pattern that produced satellite spatters. The blood in this pattern originated directly above this area of the stock with the stock lying approximately horizontal with the left side down.

Other high-velocity impact spatter on the pistol grip reflects an origin from the muzzle. A hand could not cover this area at the time of discharge.

The sweat shirt (MR1) has a bloodstain on the left sleeve at the inner left elbow area. When considering reasonable shooting positions, if the husband shot his wife, the location of this pattern was not within expectations. This stained area, as demonstrated in experiments, was typical for a transfer pattern on worn cotton cloth.

Conclusions expressed in this report may be subject to change, should new information become available.

By the time I had finished my examinations, the husband had been in jail for almost a year. The DA reviewed my findings and agreed this was a suicide. This was the only instance in my career where I was called in to evaluate the work of other opposing experts. That is unfortunate, I think this process should be more common. It is certainly more expedient and less costly than a trial.
All charges were dropped against the husband and he was released from jail.

Chapter 26 Phil Spector: The Blood and Gun Issues

Phil Spector was a noted musician, songwriter, and musical producer. He wrote several songs and produced several hit singles during the era of the Beatles, the Righteous Brothers, and other famous rock-'n'-roll groups in the 1960s and 1970s. He was especially famous for the Wall of Sound, which produced a fuller, richer sound in musical productions.

He was eventually inducted into the Rock and Roll Hall of Fame for his contributions to rock music. Since the 1970s, Phil had led a fairly reclusive life.

On February 3, 2003, Phil Spector met Lana Clarkson at a nightclub, and they eventually rode in his limo to his mansion. Both went inside while the limousine driver waited in his vehicle listening to music. The home was approximately eighty-six hundred square feet with ten bedrooms and eight and one-half baths.

Lana Clarkson was a part-time actress who began her acting career in TV shows such as *Three's Company* and *Happy Days* as well as the movie *Fast Times at Ridgeway High*.

She was also in several national commercials such as Mercedes Benz, Playtex, and Nike. At the time of the incident, she was working at the House of Blues.

The following paragraphs are a summary of what may have happened. In both trials, opposing attorneys offered slightly different renditions of this theme.

About an hour after the two of them entered the Spector home, the driver heard a gunshot, and Spector appeared in the doorway, reportedly holding a revolver. He was wearing his sport coat at the time. The driver later stated to police that he heard Spector say, "I think I killed somebody," or words close to that statement. At that time of the incident, the driver's first language was Portuguese, and the defense contended his command of the English language was poor. But years later at trial, his English had reportedly improved. So the jury got to hear a witness who was clear and concise. How do you account for that if the original statements were not recorded?
How well he understood what he heard and later communicated those words to the police was debated in both trials.

Witness recall of statements made in times of duress can be problematic. History has demonstrated many times that what was actually said may be slightly different when recalled. I have no opinion on the validity of the driver's statement; the jury believed he got it right.

As a result of the shot heard by the driver, Lana Clarkson had sustained a contact gunshot inside the mouth while sitting in a chair in a hallway near the door. After the shooting, Clarkson's head was tilted to the right, and subsequent bleeding fell on her right-side clothing and the chair.

When the police made entry, Spector reportedly resisted arrest and was Tasered, handcuffed, and later charged for the murder of Lana Clarkson. During the arrest, Spector was reportedly on the floor in front of where Clarkson was seated. This may be important when considering trace evidence.

In my review of the case, there was no controversy that Phil Spector held the revolver in his hand at some point. The weapon was found under Lana Clarkson's left leg, and it did not get there by her action. The question to be answered was, was he holding it during the shot to Lana Clarkson or afterward? If he picked it up afterward, that would not be that unusual. Remember that in the chapter "Homicide or Suicide?" the husband picked up the rifle after his wife shot herself and placed it on the bed. In this case, that would have been a really bad idea but not unusual.

At some point, it was reported that Phil Spector had obtained a cloth and attempted to wipe blood from Clarkson's head after the shooting.

He also removed his sport coat and washed his hands before police arrived. The police discovered his sport coat on the floor of his bedroom upstairs.

The scene photos were rather graphic and will not be covered here with the exception of Lana Clarkson's hands and limited bloodspatter. A portion of the artificial nail on the right thumb was missing. A white object, thought at the time to be the broken fingernail, was located on the stomach area of the deceased. This may be relevant later, as it indicated the loss occurred with this event.

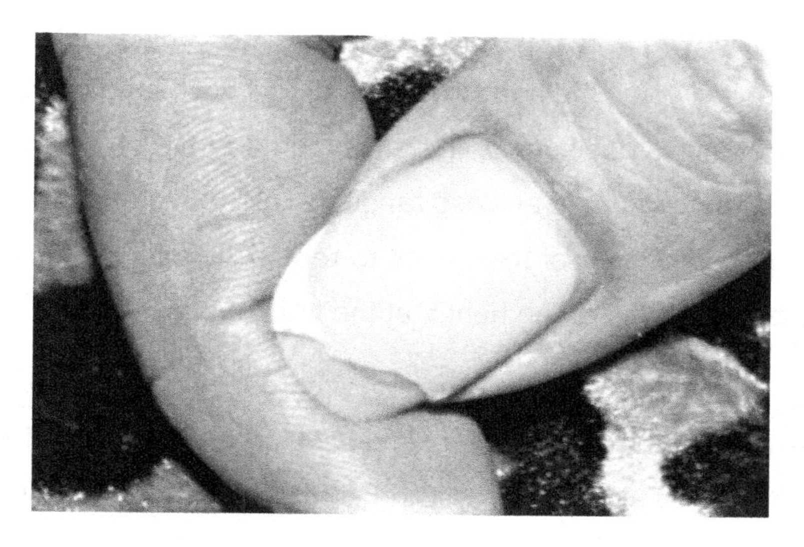

Figure 94

Based on the scene logs, I recall there were at least twenty-six people in the scene before she was removed. The scene report indicated that the coroner and his staff examined the body at the scene, including doing a UV-light exam.

This initial exam revealed no specific result. One would expect that would include looking for bloodstains on the hands since you would be able to see them with this light.

A multitude of photographs was taken during the search by the two different agencies. One agency was the crime lab, and the other was the coroner's office. However, neither agency took detailed photographs of her hands. The hands could be the best evidence available when attempting to determine the difference between a homicide and a suicide. Other than the weapon, the spatter on the hands can often tell you if the hands were holding the weapon or were in a defensive position. Or if the weapon was held by someone else. Since spatter at close range travels in a straight line, the exposed portion of the hands to receive spatter can tell you the orientation of the hands at the time of the shot.

During the first trial, the coroner's technician who examined the body at the scene said there was limited spatter on the outside wrist of Clarkson's hands despite limited photographic evidence. At the autopsy, the technician, who was not a bloodspatter expert, made some notes, "right wrist— some small apparent bloodspatter—outside wrist."

And “left wrist—outside bloodspatter” was also noted. The autopsy findings said, “ A Coroner criminalist collected (3) swabs from the wrist and mentioned possible smear/bloodspatter, though it was not described. The Medical examiner did not mention any bloodspatter on the hands.

Waiting for autopsy to look for bloodspatter is problematic, in that small spatter is a solid and the underlying skin is elastic. The bloodstains can fall off in handling and transport. Also other stains may fragment and be seen as blood flakes in other areas due to transfer. The best evidence evaluation of bloodspatter needs to be done at the scene.

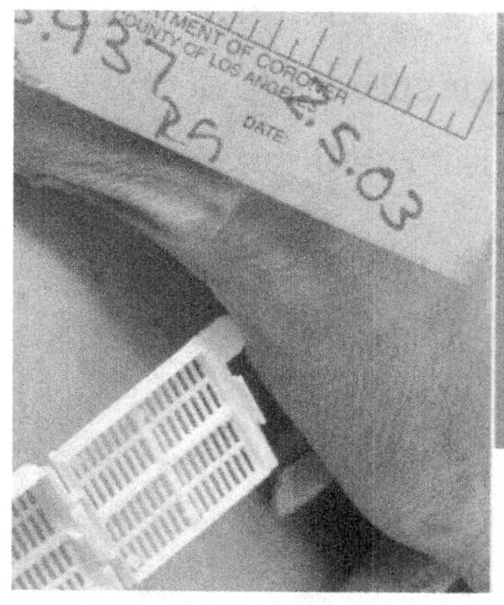

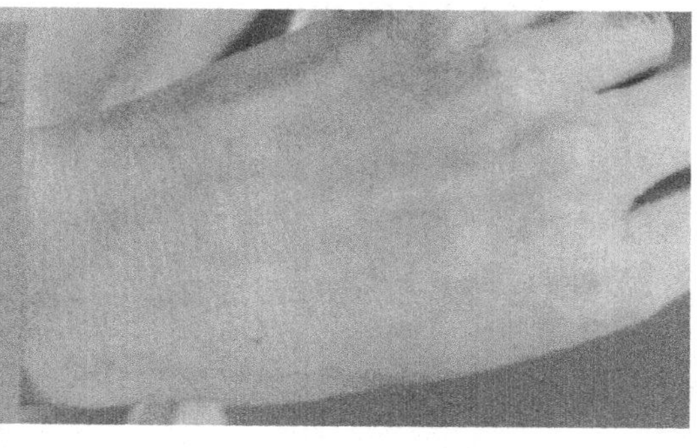

Autopsy Photos of Right Hand

Figure 95

I examined the limited autopsy color photos of Clarkson’s hands and did not see a pattern of bloodspatter on the distal wrist area of either hand, Figure 95. Finding bloodspatter on the right hand, wrist, or arm would be difficult to identify as back spatter. The arm was near the origin of satellite spatter from the chair. It is not always possible to tell satellite spatter from spatter coming from a gunshot. If both are possible, both findings must be included until one can be eliminated.

The incomplete descriptions as mentioned by the coroner's tech only confuse the issue later when both the prosecution and the defense are trying to use this limited information for their own purposes. This area of the hands in question were apparently swabbed for further evidence, so if bloodstains were present, they were lost. The photo below, Figure 96, depicts a single bloodspatter still visible on the left hand. The origin of that single drop is toward the palm side of the hand.

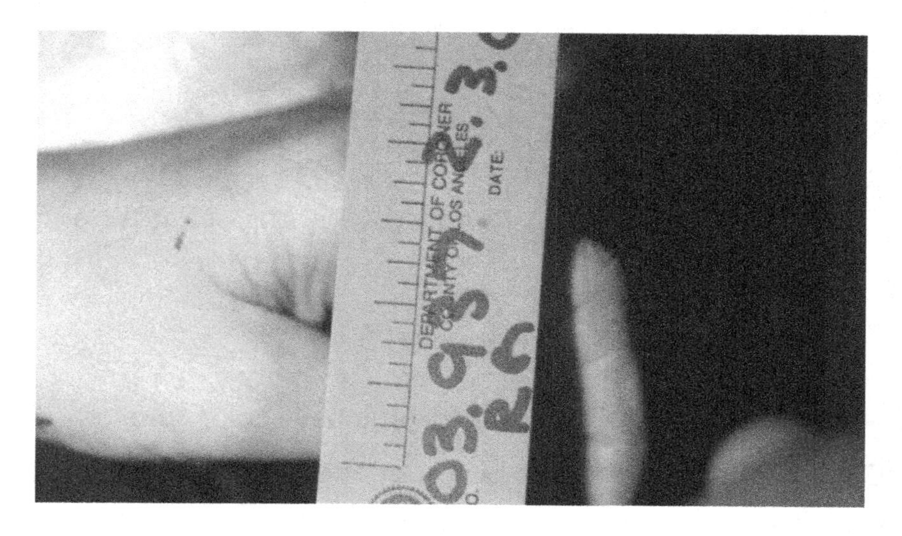

Figure 96

A review of official reports describing the scene showed that they did not mention Clarkson's hands at all, another opportunity lost. As I stated before, it cannot be overemphasized how important it is to document the hands when determining suicide versus homicide.

Simply describing an area on the hands from recall is a prescription for error. The difference of an inch could be the difference between a conviction and freedom for another person. That indeed was the case here.

Much of the bloodspatter discussed in this case was satellite spatter, back spatter, or transfer. Within the field of bloodstain-pattern analysis, the three patterns mentioned are defined as the following:

- Back spatter: A bloodstain pattern resulting from blood drops that traveled in the opposite direction of the external force applied, associated with an entrance wound created by a projectile.
- Satellite Spatter: A smaller bloodstain that originated during the formation of the parent stain as a result of blood impacting a surface. Sometimes referred to as blood dripping into blood.

- Transfer Stain: A bloodstain resulting from contact between a blood-bearing surface and another surface.

The Colt Cobra revolver used in the shooting was found by the police tucked under her leg while she remained partially upright in the chair. The weapon had obviously been placed there, as it would have been impossible to get there from Clarkson's hands after the shot. Detectives reported there was blood on both sides of the weapon, although there was no mention of blood on the carpet under the weapon. Photographs of the area did not show any blood at this location either. It is probable that Spector placed the weapon under her leg after the shooting and before the police arrived. It was also alleged that Spector washed his hands. If there was any blood on his hands, it would have been washed away.

Clarkson would remain upright in a chair for several hours while the scene was processed. During the search, detectives noted small objects on the floor thought to resemble teeth, fingernail material, or wood chips. These items were seen dispersed across the room, including the staircase. None of the small objects described on the downstairs floor was seen on Spector's jacket. If the small particulates originated from the intraoral gunshot and he was standing directly in front of her, there should have been some of this material on the jacket, his shirt, or pants.

Days later in a review of the case, the police agency decided to do a test to determine if the limo driver could actually hear a gunshot from inside the residence while sitting in his limo listening to music. Detectives went to the Spector mansion, placed a bullet trap on the floor, and had the driver and a detective sit inside the vehicle with the radio turned to music. The weapon was discharged, and both said they could hear the shot. I have a problem with the test; the shot was external to a flat surface, not inside the oral cavity of the victim. It is reasonable that there would be some noise suppression from an intraoral shot as compared to an open-air shot at the floor. But how should the shot experiment be redesigned? The muzzle should have been concealed, at least.

Later Phil Spector stated that the shooting was an accidental suicide, and he hired several forensic experts along with well-known defense attorneys to defend him after he was indicted for murder. The list of experts was a virtual who's who of the forensic sciences. A thorough investigation was conducted, and eventually Spector was tried for murder in 2006.

In the first trial, which lasted a year, an issue came up regarding impact spatter and an article that I had written in the 1980s on the subject. I received a call to testify about the possible origins of impact spatter from a firearm discharge. I traveled to Los Angeles and met with Linda Baden, the defense attorney to discuss my testimony. I was on the stand less than an hour, and the testimony was received well, as there was nothing controversial about it. I went home and assumed it was all over.

But the jury could not reach a decision on Spector's guilt, and a mistrial was eventually declared by the judge.

During this trial, an incident came up regarding my friend Henry Lee. Henry had toured the scene with several others prior to trial and was seen picking up a white object with rough edges from the floor and placing it in a vial. The vial's location was now unknown. The prosecution stated that Clarkson had lost a partial false nail and alleged that the object picked up by Lee was the nail. They made an assumption here that the nail may have been the only white object on the floor like it. Henry was accused of hiding or destroying the crucial evidence, and the nail was heralded as the evidence that would prove Spector's guilt. The prosecution believed that the nail fragment would show a trace of a passing bullet and would prove that Clarkson's hands were in a defensive position, attempting to prevent a gun being placed in her mouth.

I am not sure how that would happen, since the underlying natural nail was still there. The chance that a bullet struck the false nail and did not alter the underlying natural nail, thus proving Spector was trying to shoot her, was bewildering. If her hands were in a defensive position, her palms would be toward the weapon. A bullet might have damaged the inside of her thumb if it broke off the nail.

Then there would be the possibility of partially burned gun powder on her hands if they were between the weapon and her mouth. Damage to her teeth indicated the muzzle was inside her mouth, so her finger would also have to be inside her mouth and in front of the muzzle.

But those were the allegations at the time. Henry Lee said that he was astonished and insulted that two members of the defense team stated that he had picked up a small white object and never turned it over to the state, as the law requires. In addition, notes from the ME's office indicated that they had seen something on the floor and suggested they were possible bone fragments. Some were collected but later lost by the state. How ironic.

Not long after the mistrial, the prosecution declared that they would seek a new trial. Phil Spector hired a new attorney, and the attorney contacted me about providing testimony regarding all the blood and firearms issues in the case. I would be the only expert for the defense regarding these forensic issues. That was flattering but did not go over well with me. There had been several highly skilled experts used in the first trial, and I saw no reason to reinvent the wheel with me starting at ground zero.

My friend Stuart James had provided most of the early testimony on bloodspatter, and I advised the attorney that I would not go forward without him also being involved. He finally agreed.

Moving on to the evidence in this case, all the clothing from both Spector and Clarkson was seized and processed by the crime lab. Based on reports generated, no elemental gunshot residue (GSR) was detected on Spector's clothing. This included scanning electron microscopy of selected areas of the jacket, including the right sleeve. Spector was righthanded, and if he fired the shot, it would be reasonable there would be GSR on the right sleeve. This was significant; two trace elements are inside the priming compound of a cartridge: barium and antimony. In addition, lead and copper, elements common to the bullet, may also be detected. Barium and antimony in microscopic particulate are rare in nature, so they are commonly associated with the discharge of a firearm. The priming compound of a .38 Special cartridge contain both elements. When fired, revolvers leak a gas cloud between the cylinder and the barrel as well as out of the end of the barrel.

When this weapon is discharged, even with the end of the muzzle inside the mouth, explosion in the cartridge casts trace amounts of these elements in all directions, which will land on any adjacent objects. This cloud of trace elements was known to remain in the air after the shooting, and even if Spector was not the shooter, there was an expectation that some GSR should have been on his clothing from just being at the scene immediately after the shot was fired. Normally testing the hands and clothing of a victim of a gunshot wound was seldom worthwhile, as they would be positive for GSR anyway. It is reasonable but not certain that modern testing, such as was done in this case, would detect GSR on Spector's clothing if present. But none was found.

The clothing worn by Clarkson was laid out flat during the examination, and arrows were applied showing the direction of all bloodspatter present as represented in Figure 97. This was not an uncommon technique, but in review, one should consider using a mannequin, since the person wearing the garment was sitting upright in a chair and there were undoubtedly some wrinkles in the cloth. Spatter on wrinkled cloth could appear to be from a different direction than splatter from flat cloth. That may explain some of the arrows in the photo below.

There was bloodspatter all the way down Clarkson's outer clothing, almost to her knees. Keep in mind that the spatter observed on her clothing would be the result of either the intraoral gunshot or the satellite spatter from the chair. If the cuff of the jacket was only a few inches from Clarkson's mouth and back-spattered blood exited the mouth all the way to her knees, there should be bloodspatter on the coat cuff of the shooting hand. Multiple experts concluded there was a small single transfer stain on the cuff of the right sleeve but no spatter. However, one state crime lab examiner believed this single spot was spatter.

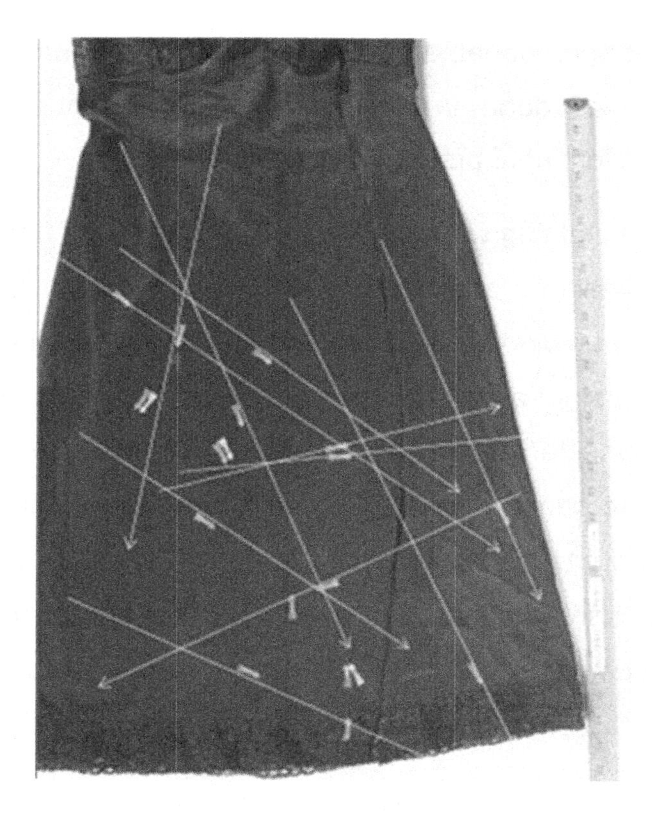

Figure 97

There was spatter on Spector's jacket but only on the left side, left sleeve, and chest right up to the midline, nothing on the right side of the jacket. How did this end at the midline so abruptly? Spector was right-handed, if the spatter was from the shot, it should be on the right side. The absence of this evidence on the right sleeve was crucial. If he pulled the trigger with the jacket on, the spatter had to be there if any spatter exited the oral cavity. With the muzzle inside the mouth as suggested by the prosecution, the distance to the cuff of the hand holding the firearm was only a few inches and in the path of any backspatter. As confirmation that backspatter was created by the shot, a report generated by the crime lab regarding the remaining exposed cartridges in the other cylinders stated,

"Some of the live rounds Item 1B were observed to have mist-like blood stains and possible soft body tissue other than blood on the rim and floor of the projectile hollow points."

The distance from these live rounds and the cuff of the jacket would only be a few inches. Stuart and I made several trips to California in pretrial conferences with the new attorney. In one trip, we visited the home of Phil Spector. This was a large mansion on a hill in Alhambra. But to my surprise, when we got there, it was in very poor condition. The grounds were unkempt and had been unkempt for quite some time.

The interior was clean but not well kept. The carpets on the main floor were a lush red and had been replaced since the incident. The walls and ceiling were off-white. In a review of the incident, we placed another chair in the hallway where the shooting took place and discussed what we knew about the case.

While we waited for Phil to present himself, we wandered around the adjacent rooms. In a large room was a guitar on the floor leaning against the wall. There was a note attached that said, "This was John's favorite, Yoko," referencing of course to John Lennon. I can't even imagine the value of that guitar.

At one point during the review, I was down on my hands and knees and looked at the chair and its position compared to photographs. While I was there, I noticed off-white flakes on the carpet. What was this? I picked up several of them and concluded that they were consistent with paint or plaster from the walls or ceiling. The particulates could be seen the full length of the hallway. Phil said that the place was not in good shape, and as an example, when it rained, the rain ran down the wall behind the chair. What also happened was that when someone walked in the hallway or slammed the door, these small but highly visible flakes of paint or plaster would fall from the ceiling. I can't imagine how much would fall from a gunshot. This was not only relevant to the case but also was most relevant to Henry Lee. What he saw and may have picked up could have been paint or plaster chips

By the time we were preparing for the second trial, there had been teams of people from both the prosecution and the defense observing the evidence and reaching conclusions. I did not have a good feeling for Phil. The quality of the experts was high, and it was unlikely that working alone, Stuart and I were going to find anything new.

I returned home and had several of the photographs enlarged and pinned them to my wall, making an oversize total photo of the garment Lana Clarkson was wearing, so that I could see the bloodspatter every day. The collage covered one wall of my office, Figure 98.

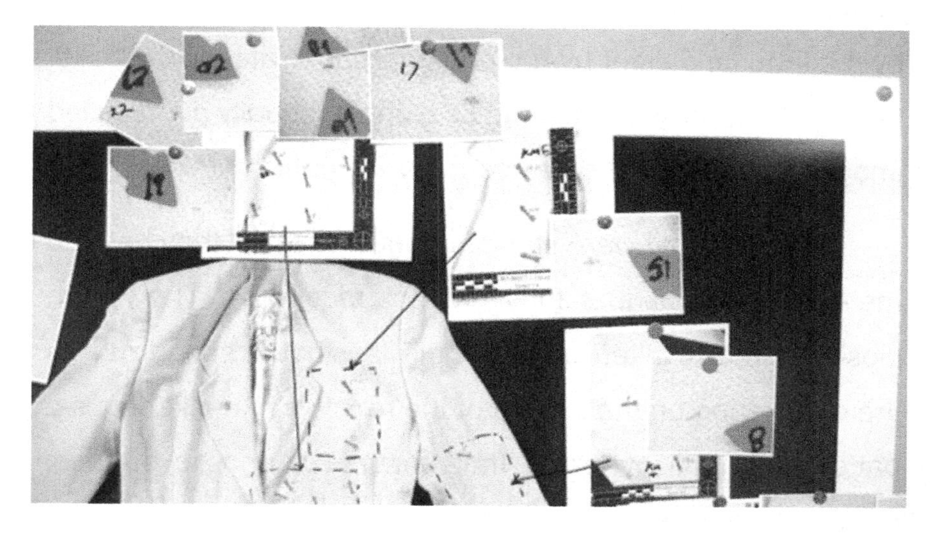

Figure 98

I did the same pinup with the chair covering another wall. The fact that the spatter had been altered before it was photographed continued to make my job very difficult. The arrows in the photographs demonstrated the location, not the direction of the spatter. Stuart and I discussed the evidence in the case and decided to split the work between us so that we could concentrate on specific items. He would take the garments Clarkson and Spector were wearing, and I would take the firearms related evidence.

Weeks later we again traveled back to California. The defense team had rented a room in an apartment building to contain all the photos and reports as sort of an office away from the courthouse. Keep in mind that every forensic person who had worked on this case had taken photographs. As a result, two of the attorneys were busy building a catalog of over a thousand eight-by-ten photographs related to the case. Occasionally they would hand us a photo, usually a magnified image, and ask what it was so that they could catalog it.

Stuart and I were standing together when they handed me some photos that they did not recognize. They were closeup photos taken by the crime lab that I recognized as the checkering on the handle of the revolver. Checkering is a pattern of small pyramids that ensures a nonslip surface. Figure 99 demonstrates how a hand covers the grip in a normal shooting position.

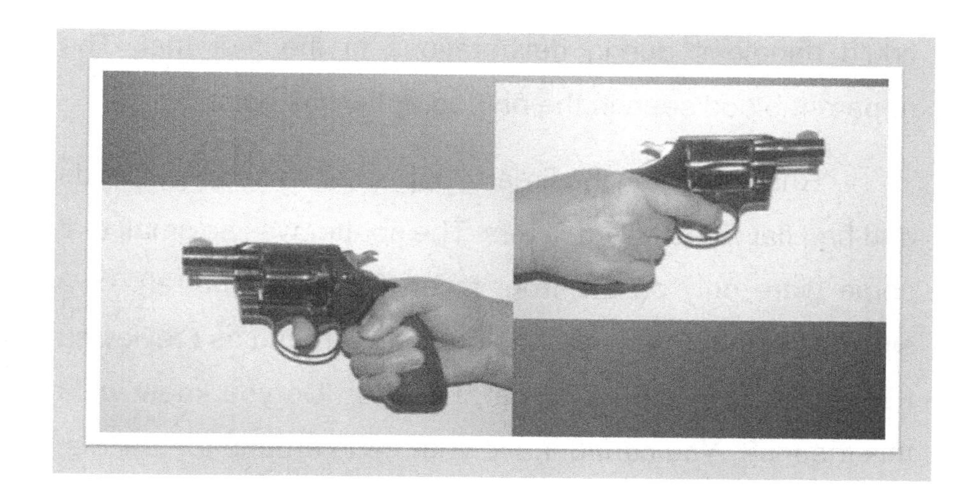

Figure 99

I was about to hand them back when it happened; I saw that there was bloodspatter on the checkering! Some of the photos were of such high magnification that I could not tell immediately the orientation of the checkering or from which direction the spatter had originated. I needed all available photos of the firearm to figure this out. Consequently, there was considerable excitement in the room. Obviously, the lab saw this too as these were the state's photos.

As I pieced the photos together, I could see that the spatter originated from the front of the weapon, the barrel end in Figure 100. There was an assumption here that the red staining interpreted as spatter seen on the firearm was blood. To be sure, I asked to see the original weapon to see for myself. I found that it had been totally cleaned so that the jury could handle it during deliberations in the first trial. The apparent blood seen in the photographs was gone.

Checkering on the firearm appeared as small pyramids and had flat sides called *facets*. The spatter was so small that some were only on the front facet (facing the barrel), of a single pyramid. Everyone was gathered around as I reviewed the photos, and then it hit me. I asked, “Do you know what this means?” You cannot have your hand around the handle, sometimes called the *grip*, and have spatter there when the weapon was discharged. How can that be? Someone had to hold it. As is common to my profession, solutions are not always immediate. There was more work to be done.

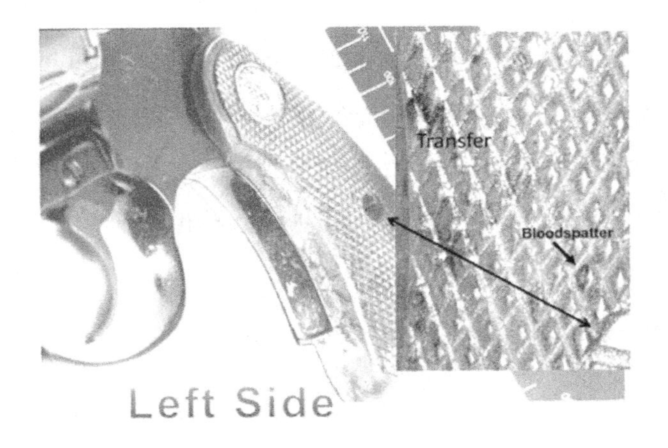

Figure 100

Prior to the first trial, all experts agreed that bloodspatter from a spatter-forming event was present on the weapon. Now the focus was on how the shooting created spatter. Two options were known: One was Clarkson bleeding onto the chair after the shot and creating secondary satellite spatter, and the other was back spatter from the contact shot in the mouth. Whatever the source, it had to be from the direction of the muzzle. Considering this dilemma from an elimination standpoint, we do not know anything about the firearm after the shot was fired. If Clarkson dropped it, it could be back spatter. If Spector fired the shot with his hand covering this area, could he have set it nearby where satellite spatter struck the grip with the muzzle up?

Figure 101 is magnified many times, it is extremely difficult to transfer blood in such small quantity in the recesses of the wood.

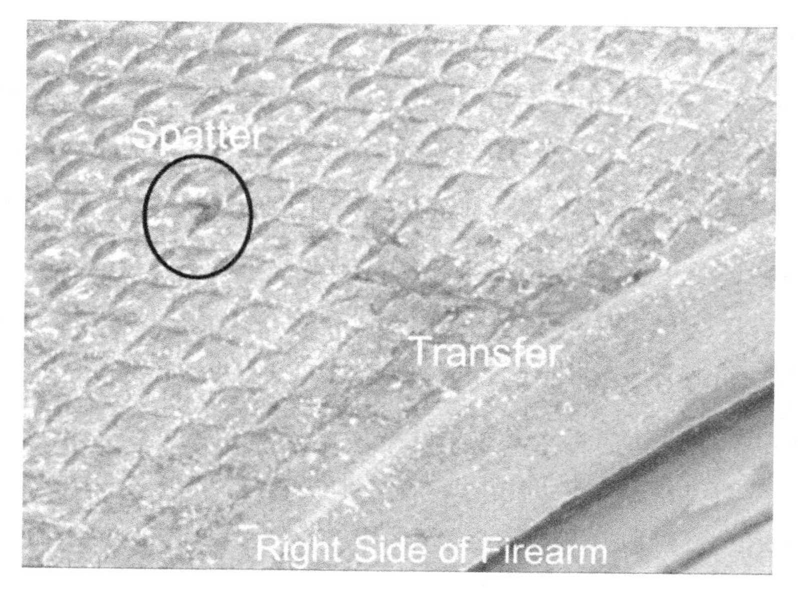

Figure 101

In addition to this spatter, there was also testimony by a prosecution expert that the weapon had been wiped, based on blood seen in the indented print on the barrel and not on the surrounding smooth surfaces. This testimony was not uncommon, and I believed this as well in my early years in this business. In doing my own experiments with blood-saturated sponges using pistols and revolvers of a variety of calibers, I had found that the blood that strikes the weapon often falls off after it dries. This loss of blood may be incomplete and most common on the smooth exposed surfaces. This gives the impression the weapon was wiped.

Blood in the indented letters are protected and may survive handling over time when the more exposed bloodstains do not.

My theory on why this happens is the failure of a dried liquid to adhere to a smooth metal surface and that most people place a light coating of oil on the metal parts of a weapon to prevent rusting. When dried, blood is a solid and easily flakes off the metal, which is facilitated by a hydrophobic coating of oil. The situation can be exacerbated by handling or placing the weapon in a loose container for storage or test firing later. The blood on the exposed surfaces falls off while the blood in the indented print is protected.

The other assumption disproved in testing was that all surfaces of the indented print would contain blood. Not true. Spatter strikes the barrel and other parts of the weapon at random and does not saturate the area, as had been expected. Again, finding some areas of indented print containing blood and the surrounding smooth surfaces devoid of blood does not mean that the weapon was wiped. Unfortunately there was testimony on this observation and it was highly inflammatory, as it suggested an effort to conceal evidence after the shooting.

The photo below, Figure 102, shows a normal hand hold and depicts blood in the medallion of the evidence weapon.

It is difficult to demonstrate in black and white photos but the tops of the small dots are absent any blood in the indented areas. In this case, it was known that the weapon was on the carpet at least once, when it was placed near Clarkson's leg. Then it was placed in an evidence container and moved to storage and the lab. Within the lab, the weapon was handled multiple times by lab personnel.

The premeditated act of intentional wiping of the barrel cannot be eliminated by microscopic exam of the indented areas. But it cannot be confirmed, either. Here at the juncture of science and law, the benefit of doubt should have gone to the defendant.

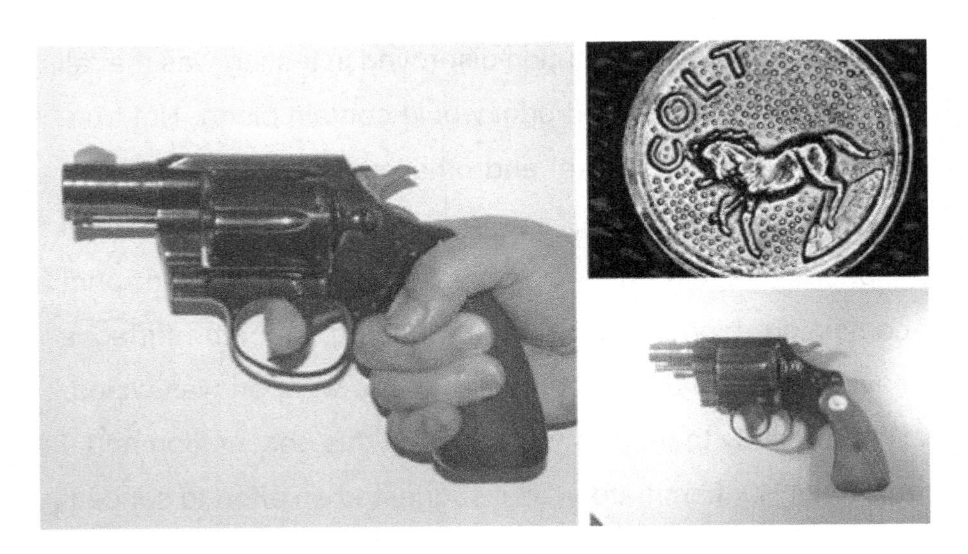

Figure 102

Go back to the photo above of the woman's hand in Figure 102, holding the colt revolver. The medallion is covered by the thumb in a normal shooting position.

If Spector held the weapon, the only way to get blood on the medallion was by transfer from his hand to the medallion after the shooting. Back spatter caused by the shot would not be the source as it would be physically blocked by the thumb. The testimony that the blood on the medallion exposed a premeditated, criminal act of attempting to conceal evidence was not there and incorrectly suggested.

Getting back to the scene, there was a three-drawer ornamental chest located next to Ms. Clarkson. The top drawer was open, and a holster fitted for the .38 revolver was inside. Apparently, no one fingerprinted the drawer or the holster, Figure 103. In review of the scene reports, I could not see any fingerprint powder or evidence of superglue, common to the fingerprint process, on the chest or the holster. Original lab notes showed that the lab swabbed the holster exterior, presumably for the presence of blood. If there were prints on the holster, the swabbing would have removed them.

Finding Clarkson's prints on the drawer or the holster might not have determined who pulled the trigger but would have been valuable information about who retrieved the weapon. In the state's defense, if this item and the ornamental chest were printed, I did not have the report. This was one of those circumstances where finding Spector's prints would mean nothing; he lived there.

But finding Clarkson's prints would have meant everything. It might have supported the contention that she retrieved the weapon.

At this point I suspect the investigators never considered the possibility that eliminating the suicide was as important as proving the murder. It was time to do some work of my own.

Figure 103

Not all of the blood visible on the firearm was spatter, as some appeared to be transfer. One must remember that a photograph represents all activity, from the incident to the time the photo was taken. There was no detailed accounting for every possible location of the firearm after the shooting regardless of who held the weapon. We do know that it was reported that Spector had a firearm in his hand when he talked to the limo driver. And the weapon did not get under Clarkson's left leg on its own. Any blood originating from Spector handling the firearm would be expected to be transfer patterns.

Back in my own lab, I had two weapons nearly identical, which I used in testing. Both had two-inch barrels, and for purposes of testing, it did not make any difference which one was used. The question was how to hold a weapon, shoot it, and not have the hand cover the area where the spatter was found. The question was not, could I prove Clarkson or Spector held the weapon? The question was, which one could I eliminate? The photos below represent two different hand positions. The hold on the right completely covers the areas where bloodspatter was seen; the hand position on the left does not. The spatter could not have landed on the checkering if the weapon was held in the usual manner shown on the right in Figure 104. This would eliminate Spector as the shooter. Clarkson could not be eliminated if the blood was back spatter.

So how to shoot a weapon in the lab holding it in a manner that would simulate the hands in the left photo, Figure 104? To shoot, I placed one finger and thumb of the left hand on top of the barrel, one finger of the right hand on the trigger, and the thumb under the hammer. With a snub-nosed .38 Special, the discharge was hard on the fingers, but I did manage to retain control of the weapon. My handhold on the weapon allowed for this area to be exposed.

As expected, the spatter found on the checkered facets faced toward the muzzle.

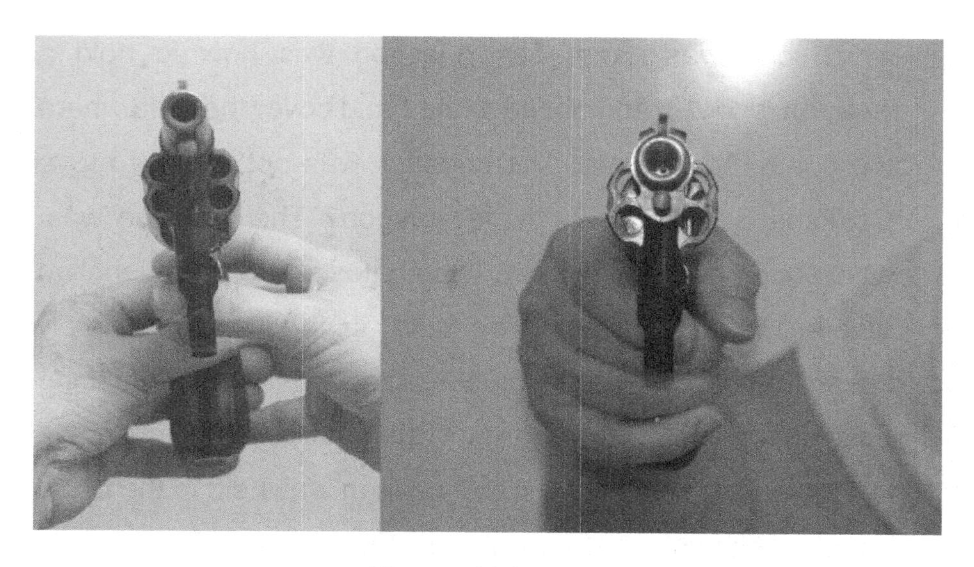

Figure 104

I had excellent color photos of this spatter but when converting them to black and white for this book, they were nearly invisible.

Time also becomes a factor here, in relation to how long it takes blood to dry. Small spatter will dry quickly, in seconds, and may not be altered by movement of the weapon. The exception is metal surfaces which have a smooth surface that causes poor adherence of blood.

Larger bloodstains may still be liquid for up to thirty minutes and subject to transfer to another object.

This transfer could be from the weapon to the floor or from a bloodstained object, such as a hand to the weapon or weapon to the hand.

Remember the broken artificial thumbnail? If the nail was broken during the incident, the graphic below in Figure 105 suggests how the nail may have been broken during the discharge of the weapon in an apparent suicide.

The force of the recoiling weapon might have broken the nail. Compare this graphic to the right thumb position in the previous graphic, Figure 104 of a female holding the weapon normally. Again this is speculation; we do not know if the break occurred during the incident. But it is a scenario that cannot be eliminated, Figure 105.

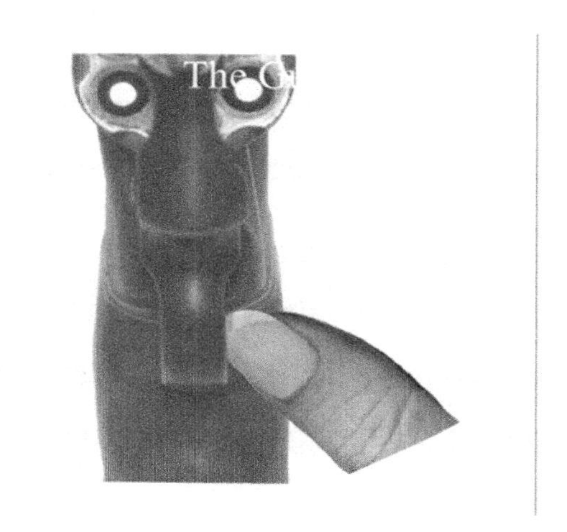

Figure 105

The prosecution suggested that her hands were around Spector's hands when he was holding the firearm; breaking the nail seemed less plausible in that scenario. All of us live in a world of truths and beliefs often reflected by our culture. In criminal investigations, it is no different; we often accept beliefs as facts. When I am reading a report, I like to separate facts from beliefs and then research the beliefs, through questioning and diligence. By this process, the truth is often revealed.

As I mentioned before, all objects, including the hands and the clothing, close to the source of spatter have the potential of receiving spatter. A report generated by three other nationally recognized experts in bloodstain-pattern interpretation indicated two possible sources of the spatter on the sport coat. One was satellite spatter from blood dripping onto and off the chair.

The other was back spatter from the shooting event. At that time both were given equal consideration.

In 1985, I published a paper in the *Journal of Forensic Sciences* titled *"Observations of High Velocity Bloodspatter on Adjacent Objects."* This was a report of testing by shooting into a blood-soaked sponge and recording the observations of the back-spattered blood. I had done hundreds of shooting tests for this paper and in subsequent casework. I know what to expect. The back-spattered blood would be deposited on the sleeve of any garment worn if it was only inches from the source.

The photo below in Figure 106 depicts experimental testing with me wearing a sport coat when shooting a blood saturated sponge with a handgun. . Bloodspatter was present on both the outside and inside of the shooting-hand sleeve of the sport coat. The amount of back spatter varies depending on the muzzle to target characteristics.

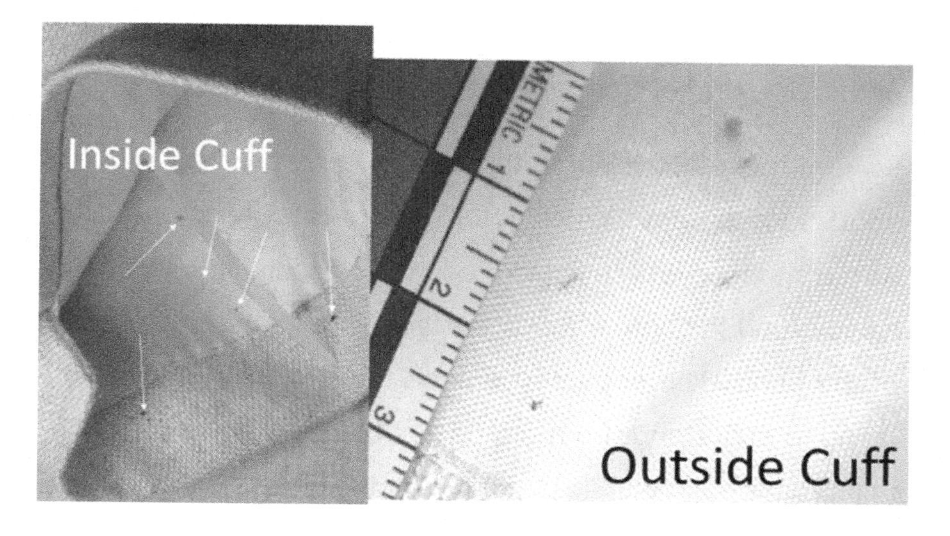

Figure 106

If Phil Spector had his jacket on at the time of the shot, if he held the firearm in either hand just outside the lips of Lana Clarkson, and we know the event created back-spattered blood, there would be small drops of blood on the sleeve of the arm holding the weapon. This would be both inside and outside the cuff. None was detected.

Regarding the sport coat, we now come back to the issue of bloodspatter only on the left side of the jacket and nothing on the cuffs. The spatter on the left side was numerous but ended at the midline seam. How does that happen? My associate, Gary Knowles, and I looked at this issue, knowing that there was earlier testimony that Spector had used a cloth to wipe Clarkson's face after the shooting. It was unknown if he was kneeling or standing.

Was the spatter a result of back spatter from the shot or satellite spatter from blood splashing on the chair?

I have talked about the bloodspatter problem with the sport coat already. When we mention satellite spatter, this was a situation where liquid blood dripped into liquid blood or liquid blood fell onto a rough surface, such as the upholstery of the chair. The impact or splash causes small droplets of blood to scatter out from the parent blood at the impact site. In this case, Clarkson lost blood after the shooting on her right side, and this blood dripped onto the arm of the chair. Satellite spatter is a result of blood dripping into liquid blood and was visible on other areas of the chair too. Repeated dripping could have caused some satellite spatter that may have struck the left side of Spector's sport coat and not the right side if he was close enough.

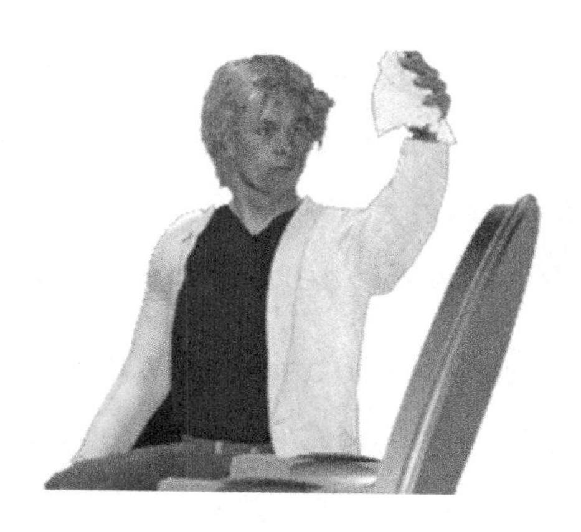

Figure 107

The graphic in Figure 107 shows how Spector could have been positioned to receive satellite spatter that struck the chair arm after the shot. The right side of the coat would not have been exposed to the spatter if it was tucked beyond and behind the right leg while he was kneeling. Keep in mind that this was not a static event as depicted in the graphic. Movement by Spector could vary the location of where the spatter struck the coat.

Phil Spector was also wearing a black long-sleeved shirt. When examined in the lab, there was a white stain on the cuffs that had properties consistent with soap. It was suspected that Spector washed his hands and may have gotten the sleeves wet at that time. If spatter was present, the washing would have removed them. If Phil Spector was not wearing the jacket in a scenario where he fired the weapon, there could have been spatter deposition on the shooting hand sleeve beyond the cuffs. No bloodstains were detected. It is important to note, the absence of evidence is not necessarily evidence by itself. The state contended that he had the jacket on at the time of the shooting, this upper area of the sleeve would be protected by the jacket.

Now we need to go back to the hands. If you consider the single spatter seen on Clarkson's left hand and the hand position as shown below in Figure 108, how that single drop of spatter was oriented to the position of the hand when the shot was fired was important. The droplet was small but long, suggesting that the blood struck the hand at a low angle. In order for blood to exit the mouth and create this drop, the hand would need to be held in a manner where the web of the thumb is close to horizontal with the mouth. The arrows point toward the origin of the spatter. How the elbows were bent would have a minor role in the exact direction of the droplet.

But how relevant is a single blood drop? Again, one must consider the time between the shooting and the photograph. After the shot, photographs show the left hand rested on the opposite side from where the satellite spatter occurred from blood dripping onto the chair.

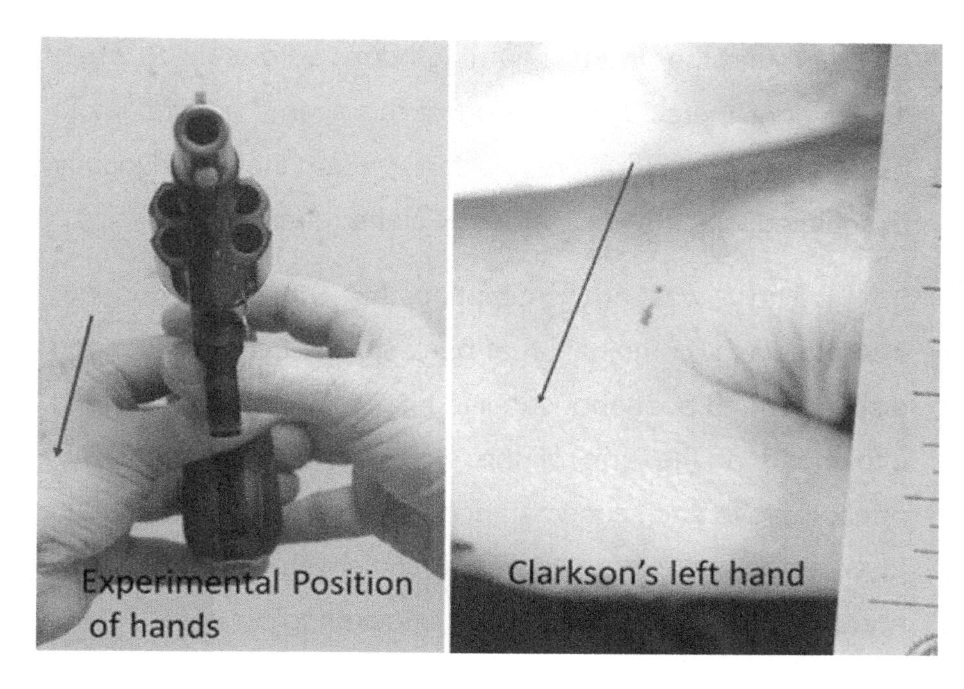

Figure 108

The chair should be eliminated as the source of this drop. If her hand was in a defensive position or her hand on his hand as Spector held the weapon, can this spatter be plausible? One would expect the angle of the droplet on her hand to be closer to ninety degrees or circular. As you read this, use your own hands, and see what you discover. As I said before, small bloodstains could fall from the skin over time as the elastic skin moved under the solid dried droplet.

Could there have been other spatter of relevance? The coroner's technician said there was. But without documentation, we will never know for sure how those stains applied to this shooting circumstance. It was unfortunate that no one found the "truth" regarding homicide or suicide when it was right there for the taking at the scene. Advocacy is always a consideration in trial, was there advocacy associated with the coroner technicians recall?

There was an assumption that the spatter on her clothing was a combination of back spatter from the weapon discharge and post-shot satellite spatter from blood dripping into blood at the arm of the chair. How much spatter is attributable to each circumstance? How does this information fit when we consider this information with what we know about the spatter on the revolver checkering?

To improve my own knowledge of oral gunshots, I contacted several other people in my profession and requested any photographs they might have of intraoral gunshots and subsequent spatter. I received photographs from about a half-dozen cases. All instances created some back spatter but not in large quantity. This supported the potential for bloodspatter external to the mouth and also confirmed by the mist-like blood on the unfired cartridges in the cylinder. This might seem like a waste of time, but in our profession, we sometimes make assumptions that seem obvious at the time. But getting complementary proof offers considerable support and is good science.

Before the second trial began, I had all of my PowerPoint presentation ready to go and discovered to the prosecution. The trial would go for almost a year before I took the stand.

In the second trial, the state provided a photograph taken from the surface of an SEM planchette. The item in the center of the photomicrograph was reported to be the cellulose backbone of a gunpowder pellet. The material on the planchette was reportedly removed from the sleeve of the Spector jacket in a process where the small circular planchette was touched to the sleeve to accumulate trace evidence. The photo in Figure 109 was reportedly magnified forty times.

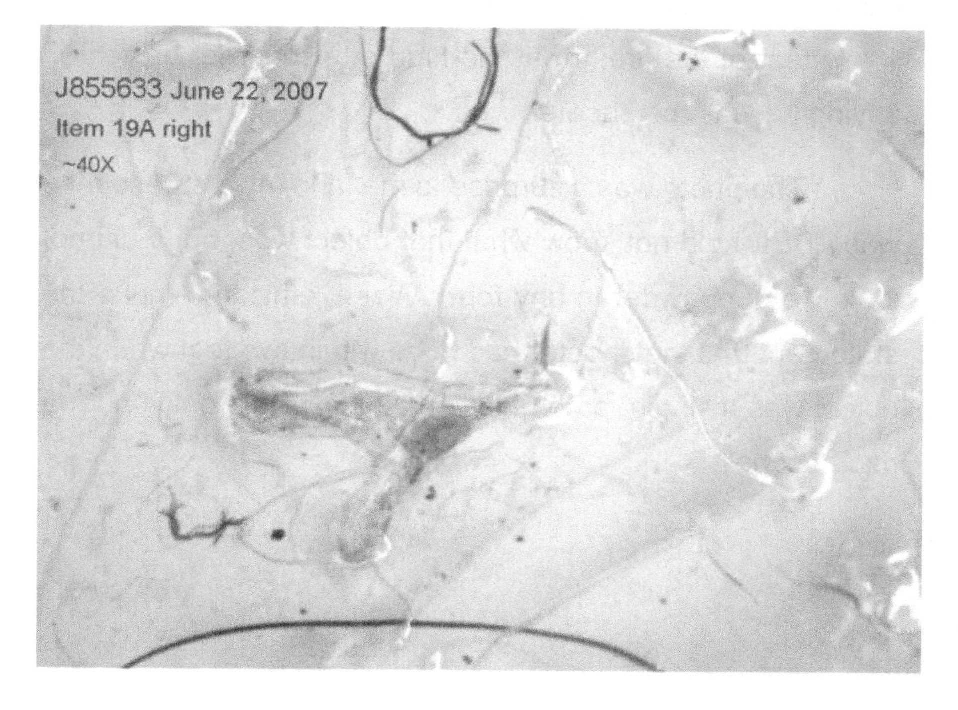

Figure 109

The significance of this object came to light when the coroner's technician's memory apparently improved from the first trial. In the first trial, I was told she remembered there was some spatter on Clarkson's hands but could not be specific as to location or quantity. Now years later and again on the stand, the technician had an epiphany; she remembered specifically where the spatter was located. It was not where it could be if Clarkson was holding the firearm. This was a direct contradiction to the blood drop in the photo above, Figure 108. With that information and the apparent cellulose-gunpowder fragment above, a state expert provided new testimony that Spector had to be within a few feet of the firearm at the time of discharge, basically saying he was the shooter.

The photo was a surprise to me, and my first response was, "BS!" I did not know what that object was, but it did not look like gunpowder in any form. Was it ethical to make the statement that the shooter had to be within two feet when the object was a single, amorphic substance visible only under a microscope?

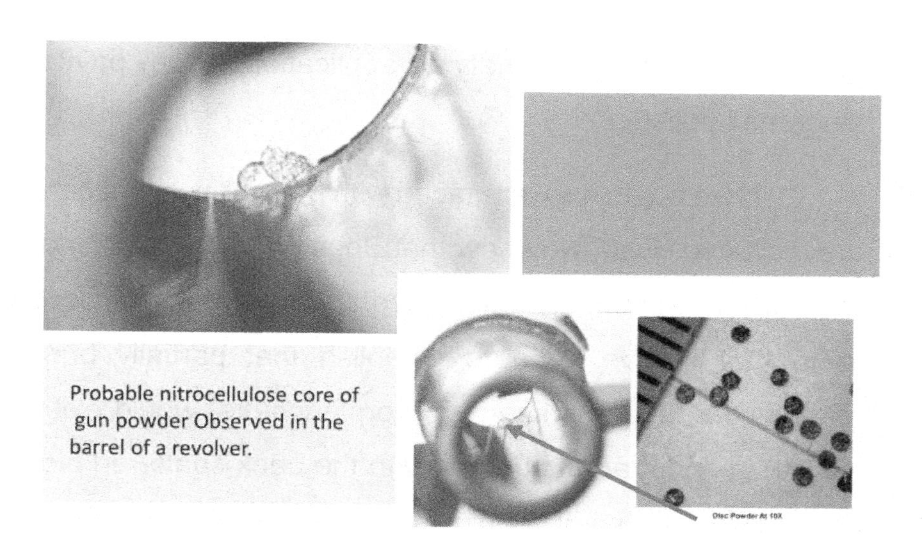

Figure 110

In Figure 110, the photo on the right with the little dark circles is the gunpowder in this case. That small circular translucent object seen in the barrel represents a probable post-discharge cellulose backbone of one of the pellets. The cellulose is usually in a circular shape and very thin.

The FBI had standards for identification of this material and required a microscopic exam to exhibit similar size and shape of the gunpowder along with the analysis for nitrates or nitrites. Neither the size or shape was similar. The additional testing for the trace elements of barium, antimony, and lead was highly recommended. Was it realistic to make the identification of gunpowder nitrocellulose, and were there other substances in the household environment that contain nitrocellulose? An Internet search showed that nitrocellulose was used in a wide variety of applications, from paints to fingernail polish.

Here again we were faced with possibilities that cannot be excluded. First, we knew that Spector was near Clarkson after the weapon discharge. Second, there was no scientific data that I have seen to establish that partially burned gunpowder, such as seen on the planchette, would exit the mouth in an intraoral gunshot with the back-spattered blood. It is a belief that it would, not science. Partially burned gunpowder is the material that causes the phenomenon known as stippling, and it penetrates or adheres to the skin or, in this case, mucosal membranes. At trial, this evidence was relevant to the state's case in combination with an undocumented recall of some bloodspatter on Lana Clarkson's wrists.

I testified that I believed the tests conducted by the state did not meet the minimum requirements for the conclusions expressed by the state.

But just for argument, let's assume the Y-shaped object was gunpowder nitrocellulose. Could it have been transferred from the dress or the chair? What about the floor in the bedroom? The nitrocellulose substance was stable. Could it be found on the jacket from some previous action? Picked up during his travels anywhere? It would depend on how long ago the jacket was cleaned. To conclude properly, what circumstances could you exclude?

In trace-evidence analysis, time is often not a factor. Solid objects may be present from previous events for a long time, and Phil Spector was known to possess several firearms. Spector was reportedly seen by the driver with the gun in his hand. Was there a transfer at that time? Could gunpowder particulates be present in the house from prior shooting events or cleaning a weapon? As stated before, it was reported that the jacket was seized by police from the floor of an upstairs room, certainly not a clean surface. Could the police also be a source of contamination? This was found to sometimes be the case in research done regarding GSR.

In my opinion, finding a single flake of theoretical "backspattered" nitrocellulose from this event on Spector's right sleeve would be highly unlikely in the absence of backspattered blood. Besides no other gunpowder or nitrocellulose was described at the scene. Yet there was no way to prove or disprove how this particle got there. In addition, no one saw any other gunpowder particulates of visible size on any of the clothing examined later, including Clarkson's clothing. The presence of unburned or partially burned gunpowder from a firearm discharge either penetrating the skin or found in the path between the muzzle and the target is well documented in the literature. I am not aware of any literature that describes these particulates from a back-spattered event, especially from an enclosed cavity such as the mouth.

It is also important to note that the particle was not consistent in shape with disc gunpowder found inside the cartridges used in this case. In light of the possibilities listed above, the prosecutorial bias by the state's expert was certainly there. One must now consider the impact of this testimony with other testimony about bloodspatter on the victim's hands that the final expert witness never personally observed or viewed in photographs.

During cross-examination, the prosecutor also asked me for an opinion about the spatter on Clarkson's hands while pointing to the distal area on his own hands as my only resource. I stated that I would not provide any testimony about bloodspatter that I have not personally seen or witnessed in quality photographs.

If he had them, I would examine them and give a response. That ended this line of questioning. There was just too much at risk for the life and liberty of another person to make a statement, no matter the circumstances or the nature of the accused. Here again is a crossroad of science and law. In the absence of certainty, the benefit of doubt should have gone to the defendant.

As can happen in high-profile cases, my own testimony was subject to an unforeseen problem. I had prepared a slide presentation to educate the jury about bloodspatter in preparation for my own testimony. In that presentation, I had shown a few photos depicting my own fingers and bloodspatter on my fingernails as an example of how spatter will land on adjacent objects with no reference to any weapon. I was asked which firearm I had used to prepare those photos, and I thought I had used the Colt Cobra. A truthful statement as that is how I had planned the presentation. I had done this testing many times with a variety of firearms, and the results were all the same. In this case, I had used both a S&W revolver with a two-inch barrel as well as the Colt Cobra with a two-inch barrel. I took several photographs of shots with both weapons. From a physics standpoint it made no difference which weapon I used. Sometime in preparing the introductory slide presentation, I inadvertently inserted slides from the S&W instead of the Colt. Months later when I was on the stand, I believed the slides were from the Colt. As it turned out, they were not.

The use of the firearms was recorded in my notes and were performed first with the S&W while I waited for the Colt to arrive. The prosecution was able to look at the properties of the digital photographs and compared them to my notes to find the connection of the photos to the weapons. Below is a photo of the two firearms. The wood grip was removed from the S&W.

Figure 111

It made no difference which firearm was represented in the slide presentation; the resulting spatter patterns were identical for both weapons. An easy way to explain this was to picture a boy with mud spattered on his pant leg.

From observing the mud, there is no way of knowing the make or size of the boots he was wearing. A contact shot, as happened in this case, is all about the physics of gases and wound ballistics. The distance between the end of the muzzle and the tissue, the expanding gases, and the opportunity for back spatter were what mattered. The make and model of the weapons involved, especially when so similar, make no difference. If you were faced with only the bloodspatter, you could not only not tell the make and model of the weapon used but also could not determine caliber.

A controversy over a few slides in an introductory slide presentation on bloodspatter was totally unexpected. But I was incorrect in my testimony, and that was the highlight of the day. I never got the chance to explain that the error was inconsequential. To make matters worse, the media made it sound like I had lied. In spite of what happened, I hold no ill feeling toward the prosecutor. He was doing his job.

In reconstruction of a scene like this one, the totality of the evidence paints a picture of what might have happened. The absence of crucial back spatter on the sport coat cuff and the presence of it on the firearm alone raise doubt that Phil Spector shot Lana Clarkson while wearing the sport coat. If indeed he did commit the crime as the jury believed, there are facts not known to me that might explain the findings. I suspect that by now we will never know.

The following was testimony given on redirect by the state's expert as revealed in the transcript:

> Q: But there are certain things in this case that you absolutely, as a scientist, can exclude that Mr. Spector

was standing across the room somewhere or in another room when Miss Clarkson was shot to death, correct?

A: Yes.

Q: You can exclude the possibility that he was more than about two or three feet away; isn't that right?

A: Yes.

Q: You can exclude it came from coughing?

A: Yes.

Q: You can exclude it came from sneezing?

A: Yes.

Q: Or CPR?

A: Yes.

Q: The only thing left was the possible source of blood— I shouldn't even use the word *possible* because now it's the only thing left; we've excluded everything else. The only thing left as the source of blood on Mr.—Mr.
Spector's jacket was what?

A: The gunshot event.

Q: And he had to be standing where?

A: Within two or three feet of the gun discharge.

The prosecution's expert witness was the only one who believed that, when four other nationally recognized experts said the single drop of blood found on the end of the left sleeve was transfer and the rest of the spatter was from the chair as seen in Figure 107. There are choices here, what can you eliminate with certainty?

Phil Spector was certainly guilty of occasional bad judgment, especially with women, and inappropriate demeanor in the courtroom. This set him up as a bad guy early in the eyes of the court spectators and the media. It certainly did not help him while the jury was reaching a verdict either. The jury concluded that Phil Spector was guilty of the charges.

Did Lana Clarkson find the weapon while sitting in the chair near the door? Did she decide to end it all? Did Phil try to intervene in the suicide attempt, placing himself close to the weapon when the discharge took place? Did he hear the shot and come to the scene afterward? Or did he shoot her? Was she attempting suicide and he was struggling over the gun with her to prevent it? Why was she in the chair near the door anyway? Was she waiting for him to arrive from another room?

Where they ready to leave? The Limo was waiting.

What was the motive if it was murder? Why shoot when there was a driver waiting just outside the door?

The decision between suicide and homicide is often difficult in the best of times. Mistakes by the state and actions by Phil himself made the issue even more complex.

We know he reportedly went to the outside doorway with the gun in his hand. The spatter on the firearm had to dry before he picked it up. How did this fit in? If it was suicide, picking up the gun and then staging the weapon under her leg was a really bad idea. As normal people, we consider what were rational actions from our own perspective. How do you judge others who are not like you? My years in this business have taught me to expect the unexpected and to seek the choices that cannot be excluded. In this case, the choices were many.

The information provided in this case was based on memory and discovery of photos and reports. My compilation of discovery documents may not be complete. Opinions expressed in this chapter are subject to change, should new information become available.

Chapter 27 Country Justice

A year before I finally retired, I received a request to look at a marijuana case from a rural county in eastern Oregon. I was asked to just look at the photos to determine if I could tell the quantity of MJ that was seized. Apparently, the owner, living on a remote property, had a medical MJ card and was allowed to have some MJ legally. At that time the law allowed for a certain number of plants under a specific size as legal for medical use. The problem was that he probably had more than the legal amount, or so it was alleged. He had plants inside his mobile home and also plants outside in a shed.

I looked at the photos taken by the sheriff's department and saw an immediate problem. There were very few photos taken inside the mobile home, and there was no scale in any of the photos. Nor were they marked in such a manner that I could tell where in the mobile home they were located. I tried to use objects, such as pots and window frames, in photos from the mobile home for scale, but I still did not see anything I could use to determine the size of the plants. All of the photos were close-ups taken for plant-identification purposes, not for size purposes.

All of the plants were removed, taken outside, and placed alongside plants that were pulled from the garden area. Now they were mixed, and there was no way to tell what were legal plants from inside or outside the residence, in the absence of a measuring device. There were more photos taken outside. As I sorted through the photos, I could not tell if some of the photos were duplicates of the same plant, either. Again, there was no scale or even an attempt to weigh any of the plants. After the photos were taken, all the MJ was picked up and destroyed.

My job, if I could do it, was to determine what he was raising within the confines of his MJ medical card and what plants were outside the allowed allocation. This was not possible based on the method of collection and the limited photos taken. I called the defense attorney and advised him of my findings, and the case was set for trial. I had hoped this one would go away, as it was a two-day drive to get to the county seat for the trial. I had worked a few cases in this rural county in the early years of my career and had got to know some of the folks in the DA's office. The judge in this case was the DA way back when. The current DA was an old friend who used to be in a DA's office on the coast, and we had worked many cases together. I looked forward to seeing both of them again.

After our arrival and just before trial, my wife and I met with the young defense attorney in a small crowded office with other folks moving about nearby. Privacy was moment by moment. This was an indigent defense case. Often young lawyers work these cases for the experience, as the pay is less than private casework. It was apparent that the lawyer had very little experience, but he was more than willing to listen to me and anxious to jump into the courtroom for trial. We talked about how best to present the evidence and what photographs to use. Finally, we went to lunch, as trial was scheduled for one thirty.

After lunch, I drove to the courthouse and looked up the DA. We were happy to see one another, as we had not talked in many years. We talked about fun times with old cases, ones that were appealed, and the way our other friends were doing. As the time drew close, we walked to the court room for trial. The judge came in the door in plain clothes, and we had a good welcoming for two people who had not seen each other in over twenty years. The DA, judge, and I reminisced over old cases from a time when we were young and excited about our work. When the defense attorney arrived, the three of us were sitting in the gallery, just laughing and talking.

His eyes projected a moment of worry. I could tell he was asking himself what was going on. The judge stepped out and put on his robe, and when he returned, the trial was in session. We had a problem; the time had come and gone for the defendant to be present, and he was nowhere to be found.

"Please be seated," the judge said.

The judge and the rest of us made small talk and waited a little longer. There was talk of having the trial without the defendant, when suddenly the defense counsel realized that he had given the wrong date to the defendant, so the defendant did not know this was his trial time. With shaking hands, the distraught defense attorney attempted to reach his client by phone just in case he was nearby. The judge was giving him a stern look of displeasure.

The defense attorney called the defendant on his cell phone and made contact. He was with a couple of friends in an old pickup up in the mountains where there was limited cellphone service. They stopped the truck while they had some cell-phone service, and the judge asked the defendant if he was willing to proceed, since he was too far away to get back any time soon. The judge did not want to spend tax-payer dollars to bring me back again. The defendant agreed, so somewhere in the mountains of southeastern Oregon, the defendant raised his right hand and was sworn in for pretrial motions to tell the truth, the whole truth, and nothing but the truth, so help him God. After preliminary motions, opening statements were heard over his cell phone as the state began to put on their case. At issue was the possession of too many marijuana plants for his medical card, too many and too large.

The client's friends got tired of sitting by the roadside, so they kicked the client out of the vehicle, drove off, and left him standing there. The defendant was some fifty miles from the courthouse swatting flies and mosquitoes on a hot summer day while we were to proceed via speakerphone. I could hardly hold back the laughter as I thought about this guy somewhere in the wilderness hearing his future decided, hopefully before he was eaten alive by bugs or ran out of battery.

I was called to the witness stand to talk about the MJ handling and the way it should have been done, and I tried to do my job for the defense. The sheriff's office had done a poor job of documenting the plants, and it was difficult to determine just how many the defendant had and just how large they were prior to destruction. On the legal side, this was kind of a David and Goliath case, only the DA, a.k.a. Goliath, was tying the young defense attorney up in knots at every opportunity, and I could tell he was out to have some fun. Keep in mind, there was order in the court, but if you had something to say, you just said it. It was kind of a casual but orderly conversation.

The judge turned to me and asked a few questions after direct and cross, and now I was involved in the total discussion of deciding the client's future while remaining on the stand. Since his arrest, the state had legalized recreational MJ, and we were discussing how to apply the new rules to the old case. I was again trying to visualize this guy at some mountainous location on a gravel road, all by himself, listening to this conversation with only the deer and the elk as witnesses. His freedom was in the balance. I was smiling about the whole affair, but my wife, who was seated in the gallery, was beside herself; she had never seen country justice in action.

As the trial progressed, I was entertained by the tenured DA chewing up the other attorney, who was a nervous wreck. The case dragged on, and the battery was getting low on the client's cell phone. Picture the four of us as we continued to discuss the client's fate while he was still sitting next to a tree on a gravel road about to lose any further communication. If he was later found guilty, I guess someone from the sheriff's office would have to go out and find him. The court discussed postponing the trial and the judge asked me about my availability for coming back. He listed specific dates and I said I would have to check with my secretary. I looked at my wife and everyone turned around to see her response. She said, "no".

At that point, the judge suggested a settlement. This was bantered about for a few minutes, and all agreed to something about time served. It seemed that he was initially unable to make bail and got to enjoy three hots and a cot for a while. Apparently, it was long enough, and such was the judge's final ruling. I smiled and had to pinch myself, believing that I was actually getting paid for this fun.

We inquired over the speakerphone as to how the client was going to get home. He did not know, other than by the shoes he had on and depending on how much battery he had left in his cell phone to reach someone.

After the gavel fell, the judge asked if I had played their new golf course in town and if I had the time to play. The young defense attorney was again horrified. While still sitting on the stand, I told the judge that the distance from home was too great; I was going to pass and needed to be on my way. I promised I would return in the near future, and we would get a game in. Over the years, my wife has sat through many trials but had never seen anything like this. During this phase of the trial, she was occasionally making eye contact with me and shaking her head in disbelief.

I believe justice was served but in a manner not often seen today, and it will probably never will be seen by my wife and I again. Despite the casual manner of trial, and the lack of sophisticated scientific inquiry, the result was fair.

The trip home was uneventful, and my wife and I had a good laugh about the experience. This was so much fun. After all of the recent advances in forensic science and all the years of my education and experience, it was just good old country justice that prevailed.

Acknowledgements

The first person that needs acknowledgement and thanks is my wife, Sally. My partner of nearly fifty years. Her support and love carried through both the good times and the bad. Both of us care dearly for our children and now our grandchildren. I hope they get some enjoyment reading the exploits of their grandfather.

I was fortunate in that my division director, Major Roger Dingeman, and my lab director and mentor, Charles Vaughan, gave me opportunity and direction when I was new to the crime lab and had my back when times were tough. I owe a debt of gratitude to both of them as good friends and leaders in the profession. Through them I was able to spread my wings.

There were several people in my early career who were professional scientists and were role models to me. Within the Eugene lab, this included Mike Hurley, Terry Bekkedahl, Linton Von Beroldingen, Kenn Meneely, and Gordon Rutter as outstanding examples of quality people willing to go the extra mile when called upon. Mike Howard, Jon Spilker, Gary Knowles, and Rocky Mink were also special people with a dedication to seeking the truth as well as friends to me. They had an impact on my career, and I hold all of them in the highest regard. As police officers and scientists, their ethical standards and quality of work were beyond reproach.

Within the Coos Bay crime lab, my staff members Cheryl Waddington, Gail Sampson, and Linda Osterhaus were professional individuals who offered support and friendship for many years. We were as close as co-workers could be in a demanding environment. I looked out for them and they looked out for me. God Bless them all.

A special thanks goes out to David Pex, my cousin, for his skills with Word and his editorial review prior to publishing this book.

Made in the USA
Columbia, SC
30 September 2019